Business Case Analysis with R

Simulation Tutorials to Support Complex Business Decisions

Robert D. Brown III

Apress®

Business Case Analysis with R: Simulation Tutorials to Support Complex
Business Decisions

Robert D. Brown III
Cumming, Georgia, USA

ISBN-13 (pbk): 978-1-4842-3494-5   ISBN-13 (electronic): 978-1-4842-3495-2
https://doi.org/10.1007/978-1-4842-3495-2

Library of Congress Control Number: 2018935875

Managing Director, Apress Media LLC: Welmoed Spahr
Acquisitions Editor: Susan McDermott
Development Editor: Laura Berendson
Coordinating Editor: Rita Fernando

Cover designed by eStudioCalamar

Cover image designed by Freepik (www.freepik.com)

Distributed to the book trade worldwide by Springer Science+Business Media New York, 233 Spring Street, 6th Floor, New York, NY 10013. Phone 1-800-SPRINGER, fax (201) 348-4505, e-mail orders-ny@springer-sbm.com, or visit www.springeronline.com. Apress Media, LLC is a California LLC and the sole member (owner) is Springer Science + Business Media Finance Inc (SSBM Finance Inc). SSBM Finance Inc is a **Delaware** corporation.

For information on translations, please e-mail rights@apress.com, or visit http://www.apress.com/rights-permissions.

Apress titles may be purchased in bulk for academic, corporate, or promotional use. eBook versions and licenses are also available for most titles. For more information, reference our Print and eBook Bulk Sales web page at http://www.apress.com/bulk-sales.

Any source code or other supplementary material referenced by the author in this book is available to readers on GitHub via the book's product page, located at www.apress.com/9781484234945. For more detailed information, please visit http://www.apress.com/source-code.

Printed on acid-free paper

To Jan, who has provided unyielding support
for my various crazy projects over the years, especially
when I said, "I'm going to write a book."

And, to my parents, who first instilled in me a
lifelong pursuit of learning.

I love you all.

Table of Contents

About the Author

Robert D. Brown III is the President of Incite! Decision Technologies LLC, a consultancy supporting senior decision makers facing complex, high-risk opportunities. These opportunities usually include strategic planning, project selection, planning and risk management, and project portfolio analysis and management.

Mr. Brown has devoted his 20-year career to providing solutions to his clients' complex problems by employing creative thinking and advanced quantitative business, engineering, and systems analysis. His client experience spans diverse industrial and commercial fields including petroleum and chemicals, energy, utilities, logistics and transportation, pharmaceuticals, electronics manufacturing, telecommunications, IT, commercial real estate, federal agencies, and education.

Through Incite!, Mr. Brown delivers analysis, decision support tools and systems, and training in decision making and risk management. His goal is to help people measure the value and the risk associated with the important decisions they face to make informed trade-offs and choices.

Mr. Brown graduated from the Georgia Institute of Technology in 1992 with a bachelor's degree in mechanical engineering (Co-op program).

About the Technical Reviewer

Dwight Barry is a Principal Data Scientist at Seattle Children's Hospital in Seattle, Washington. He has worked in analytics for more than 20 years, in both academia and industry, in the fields of health care, hospital administration, clinical and hospital care, environmental science, and emergency management.

Acknowledgments

This book would never have come to fruition had it not been for the Atlanta R User's Group led by Derek Norton of Microsoft (formerly Revolution Analytics). They were the first group to hear my pitch for the ideas I had about how R might be used in a different way than typically employed by data scientists. Derek especially provided some key insights into the syntax of R and encouraged me to expand my ideas into what eventually became the first distributed precursor to this book.

Next, a big thank you goes to Dwight Barry. Dwight was one of the original readers of the first version of this book, giving it a glowing review that continues to warm my heart today. Being a fellow code monkey (author of his own *Business Intelligence with R* [Leanpub, 2016]), Dwight graciously offered to act as the technical reviewer on this project, providing an expert eye for cleaning up my code and tightening some of the language describing complex ideas. Of course, any ambiguity or clumsiness that remains is entirely due to my own failings as an author. Although he required little incentive from me to provide the technical review of this book, Dwight still did so as a distraction from his very important work with the data science group at the Seattle Children's Hospital and from his family over the 2017 winter holidays. Dwight is a code monkey with a heart.

Finally, much appreciation goes to the team at Apress. In the process of publishing this work, they continued to demonstrate the greatest level of enthusiasm and professional patience as I attempted to meet project deadlines (and sometimes failed) while running my own business during the day. Thank you, Susan McDermott, Rita Fernando, Laura Berendson, and Teresa Horton!

Introduction

Welcome to **Business Case Analysis with R: Simulation Tutorials to Support Complex Business Decisions**. This book first appeared as a series of four short, self-published tutorials conceived to provide specific guidance in tricky areas associated with the delivery process of project business case and risk analysis. Each tutorial was intended to be read in one to three hours, spanning no more than 100 pages so that the reader can understand big ideas quickly on a first pass. With this publication, we bring those tutorials together in a more integrated fashion as a single volume of four parts.

Although each of the four parts can still be read independently, I recommend taking them in the order of their presentation in the text. Part 1 represents the namesake of the current volume, "Business Case Analysis with R: Simulate Complex Business Decisions with Greater Transparency." This is a tutorial for learning how to use the statistical programming language R to develop a business case simulation and analysis with greater transparency, efficiency, and accuracy than is possible with spreadsheets.

The tutorial follows the case in which a chemical manufacturing company considers constructing a chemical reactor and production facility to bring a new compound to market. There are numerous uncertainties and risks involved, including the possibility that a competitor will bring a similar product online. The company must determine the value of making the decision to move forward and where they might prioritize their attention to make a more informed and robust decision. Although the example used is a chemical company, the analysis structure it presents can be applied to just about any business decision, from IT projects to new product development to commercial real estate. In this section, you will learn how to do the following:

1. Set up a business case abstraction.

2. Model the inherent uncertainties in the problem with Monte Carlo simulation.

3. Develop sensitivity analysis that tells you which uncertain effects matter most.

4. Communicate the results graphically.

5. Draw appropriate insights from the results.

Business case analysis of a new opportunity or problem solution is never complete, however, unless we answer this question: "Compared to what?" Answering this question is the business of opportunity cost analysis. Unfortunately, when people even remember to do opportunity cost analysis, they set it up incorrectly in either one of two ways.

First, they might consider a "mostly best" solution to their problem or opportunity, then they try to account for the variation it might experience by applying different levels of scenarios (i.e., low, most likely, and high) to key uncertainties. There are two problems with this approach:

- It doesn't account for the probability that any of those scenarios will occur.

- It treats the opportunity costs as arising from the existential variation on a given choice as opposed to considering the net conditional value between purposefully distinct choices.

Opportunity cost is about the trade-off in value for what we choose to do as opposed to what can happen to us in any given choice.

Second, they might try to consider the pros and cons of individual decision alternatives, but they soon face the realization that business decisions are really composed of multiple coordinated decision alternatives in which the possible combinations of alternatives approach hundreds if not thousands, nor can each alternative be treated independently. The confusion that arises from this situation usually propels the analysis back to the first approach: Pick a "mostly best" solution and try to account for possible scenario variations.

The purpose, then, of Part 2, "It's Your Move: Create Valuable Strategic Decisions When You Don't Know What to Do," is to show you how to simplify this thought process by using three thinking devices called the decision hierarchy, the strategy table, and a qualitative description table to frame creative decision strategies that effectively reduce the decision complexity of business case analysis. It gives you the ability to create the right combinations of decision alternatives for opportunity cost analysis without getting mired in testing all of the possible combinations or simplistically guessing at the best pathway to take at too high of a level of consideration. This sets the stage for conducting opportunity cost analysis in the right way by answering this question: What is the value of doing A versus B, C, or D?

Of course, we must still account for the existential variation that we can experience in any strategic pathway we choose. We need to give consideration to the effects of outcomes we cannot control and their relative likelihood of occurring. This feels like it should be satisfied through the statistical analysis of data. Unfortunately, here again, business analysts often face a significant problem: Where does one get the data, especially if the decision problems at hand merit new-to-the-world solutions and the solutions would be counterfactuals to each other? The answer is that we construct the data from subjective expert guidance via an objective process. Part 3, "Subject Matter Expert Elicitation Guide: Assess Uncertainties When You Don't Have (Much) Data or a Clairvoyant," provides a facilitation framework for overcoming many of the spoilers to clear thinking while providing a means to account rationally for uncertainty in business case analysis. It will help you do the following:

1. Recognize appropriate subject matter experts (SMEs) to use in assessing information about uncertain events.

2. Provide you with a kind of script for helping SMEs acknowledge and assess their own bias.

3. Provide you with a routine to elicit the appropriate information from SMEs required to support high-quality business case analysis.

This section provides the basis for the characterization of uncertainty employed in Part 1 and finally in Part 4.

Opportunity cost analysis done correctly often leads to the realization that several material uncertainties can pose a significant possibility that we will experience regret for taking the best indicated decision strategy over the next best one given the information we have at hand (often constructed from expert guidance). We need a tool to determine just how much we should budget to improve the quality of the information we possess about the uncertainties and prioritize our attention on them. Otherwise, we run the risk of either incurring too much exposure or spending too much to achieve the confidence that we have chosen well. Part 4, "Information Espresso: Use Value of Information to Make Clear Decisions Efficiently," provides just such needed guidance by returning us to the roots of the title of this volume, using the R programming language to achieve greater transparency into the value of the decisions we want to make. In this section, you will learn the following:

1. The meaning of value of information.

2. How to identify critical uncertainties using tornado charts.

3. How to calculate the value of information using R.

You can download the source code from this book's product page at www.apress.com/9781484234945 by clicking the source code button there. If you want to read this source code in a noncomputing medium (i.e., ebook or print book), Appendices A through E also contain the uninterrupted R source code and SME elicitation instructions.

My hope is that by using these tools you can accelerate and improve the value you bring to your own career as an analyst, and by doing so, create value for everyone who benefits from your improved skill and acumen. If you find that you enjoy and benefit from this book, please help spread the word by tweeting #BizSimWithR and include me on the tweet at @InciteDecisions.

PART 1

Business Case Analysis with R

Simulate complex business decisions with greater transparency

CHAPTER 1

A Relief from Spreadsheet Misery

Business case analyses that are typically developed in spreadsheets are fraught with a lack of transparency and prone to propagating significant coding errors. The R programming language provides a better alternative for creating clear and minimal-error analysis.

Why Use R for Business Case Analysis?

Even if you are new to R, you most likely have noticed that R is used almost exclusively for statistical analysis, as it's described by The R Project for Statistical Computing.[1] Most people who use R do not frequently employ it for the type of inquiry that business case analysts use spreadsheets to select projects to implement, make capital allocation decisions, or justify strategic pursuits. The statistical analysis from R might inform those decisions, but most business case analysts don't employ R for those types of activities.

Obviously, as the heading of this section suggests, I am recommending a different approach from the status quo. I'm not just suggesting that R might be a useful replacement for spreadsheets; rather, I'm suggesting that better alternatives to spreadsheets be found for doing business case analysis. I think R is a great candidate. Before I explain why, let me explain why I don't like spreadsheets.

Think about how spreadsheets communicate information. They essentially use three layers of presentation:

1. Tabulation

2. Formulation

3. Logic

---

[1]http://www.r-project.org

© Robert D. Brown III 2018
R. D. Brown III, *Business Case Analysis with R*, https://doi.org/10.1007/978-1-4842-3495-2_1

When we open a spreadsheet, usually the first thing we see are tables and tables of numbers. The tables might have explanatory column and row headers. The cells might have descriptive comments inserted to provide some deeper explanation. Failure to provide these explanatory clues represents more a failing of the spreadsheet developer's communication abilities than a failing of the spreadsheet environment, but even with the best of explanations, the pattern that emerges from the values in the cells can be difficult to discern. Fortunately, spreadsheet developers can supply graphs of the results, but even those can be misleading chart junk. Even when charts are well constructed, their placement in models often doesn't clearly indicate which array of values is being graphed, presenting an exercise for the user to determine.

To understand how the numbers arise, we might ask about the formulas. By clicking in a cell we can see the formulas used, but unfortunately the situation here is even worse than the prior level of presentation of tables of featureless numbers. Here, we don't see formulas written in a form that reveals underlying meaning; rather, we see formulas constructed by pointing to other cell locations on the sheet. We do not see easily how intermediate calculations relate to other intermediate calculations. As such, spreadsheet formulation is inherently tied to the structural layout of the spreadsheet, not necessarily one that reveals the inherent relationship of the ideas it encodes. This is like saying that the meaning within a book is related to its placement on a bookshelf, not the development of the ideas it contains.

Although the goal of good analysis should not be more complex models, a deeper inquiry into a subject usually does create a need for some level of complexity that exceeds the simplistic. As a spreadsheet grows in complexity, though, it becomes increasingly difficult to extend the size of tables (both by length of indexes that structure them and the number of indexes used to configure the dimensionality) as a direct consequence of its current configuration. Furthermore, if we need to add new tables, choosing where to place them and how to configure them also depends almost entirely on the placement and configuration of previously constructed tables. So, as the complexity of a spreadsheet increases, it naturally leads to less flexibility in the way the model can be represented. It becomes crystallized by the development of its own real estate.

The cell referencing formulation method also increases the likelihood of error propagation, because formulas are generally written in a granular manner that requires the formula to be written across every element in at least one index of a table's organizing structure. Usually, the first instance of a required formula is written within one cell in the table; it is then copied to all the appropriate adjacent cells. If the first formula is incorrect, all the copies will be wrong, too. If the formula is sufficiently long and complex, reading it to properly debug it becomes very difficult. Really, the formula doesn't have to be that complicated or the model that complex for this kind of failure to occur, as the recent London Whale VaR model[2] and Reinhart-Rogoff Study On Debt[3] debacles demonstrated.[4] Of course, many of these problems can be overcome by analysts agreeing on a quality and style convention. Even though several of these conventions are available for reuse, they are seldom employed in a consistent manner (if at all) within an organization, and certainly not across similar commercial and academic environments.

All of this builds to the most important failure of spreadsheets–the failure to clearly communicate the underlying meaning and logic of the analytic model. The first layer visually presents the numbers, but the patterns in them are difficult to discern unless good graphical representations are employed with clear references back to the data used to construct them. The second layer, which is only visible if requested, uses an arcane formulation language that seems inherently unrelated to the actual nature of the analysis and the internal concepts that link inputs to outputs. The final layer–the logic, the meaning, the essence of the model–is left almost entirely to the inference capability of any user, other than the developer, who happens to need to use the model. The most important layer is the most ambiguous, the least obvious. I think the order should be the exact opposite.

When I bring up these complaints, the first response I usually get is, "Rob! Can't we just eat our dinner without you complaining about spreadsheets again?" When my dinner company tends to look more like fellow analysts, though, I get, "So what? Spreadsheets are cheap and ubiquitous. Everyone has one, and just about anyone can figure out how to put numbers in them. I can give my analysis to anyone, and anyone can open it up and read it."

Free, ubiquitous, and easy to use are all great characteristics of some things in their proper context, but they aren't characteristics that are necessarily universally beneficial for decision aiding systems, especially for organizations in which complex ideas are

---

[2]http://www.businessinsider.com/excel-partly-to-blame-for-trading-loss-2013-2

[3]http://www.businessinsider.com/thomas-herndon-michael-ash-and-robert-pollin-on-reinhart-and-rogoff-2013-4?goback=%2Egde_1521427_member_234073689

[4]You will find other examples of spreadsheet errors at Raymond Panko's web site at http://panko.shidler.hawaii.edu/SSR/index.htm. Panko researches the cause and prevalence of spreadsheet errors.

formulated, tested, revisited, communicated, and refactored for later use. Why? Because those three characteristics aren't the attributes that create and transfer value. Free, ubiquitous, and easy to use might have value, but the real value comes from the way in which logic is clearly constructed, communicated, stress tested, and controlled for errors.

I know that what most people have in mind with the common response I receive are the low cost of entry to the use of spreadsheets and the relative ease of use for creating reports (for which I think spreadsheets are excellent, by the way). Considering the shortcomings and failures of spreadsheets based on the persistent errors I've seen in client spreadsheets and the humiliating ones I've created, I think the price of cheap is too high. The answer to the first part of their objection–spreadsheets are cheap–is that R is free; freer, in fact, than spreadsheets. In some sense, it's even easier to use because the formulation layer can be written directly in a simple text file without intermediate development environments. Of course, R is not ubiquitous, but it is freely available on the Internet to download and install for immediate use.

Unlike spreadsheets, R is a programming language with the built-in capacity to operate over arrays as if they were whole objects, a feature that demolishes any justification for the granular cell-referencing syntax of spreadsheets. Consider the following example.

Suppose we want to model a simple parabola over the interval (–10, 10). In R, we might start by defining an index we call x.axis as an integer series.

```
x.axis <- -10:10
```

which looks like this,

```
[1] -10 -9 -8 -7 -6 -5 -4 -3 -2 -1 0 1 2 3 4 5 6 7 8 9 10
```

when we call x.axis.

To define a simple parabola, we then write a formula that we might define as

```
parabola <- x.axis^2
```

which produces, as you might now expect, a series that looks like this:

```
[1] 100 81 64 49 36 25 16 9 4 1 0 1 4 9 16 25 36 49 64 81 100.
```

Producing this result in R required exactly two formulas. A typical spreadsheet that replicates this same example requires manually typing in 21 numbers and then 21 formulas, each pointing to the particular value in the series we represented with x.axis.

The spreadsheet version produces 42 opportunities for error. Even if we use a formula to create the spreadsheet analog of the x.axis values, the number of opportunities for failure remains the same.

Extending the range of parabola requires little more than changing the parameters in the x.axis definition. No additional formulas need to be written, which is not the case if we needed to extend the same calculation in our spreadsheet. There, more formulas need to be written, and the number of potential opportunities for error continues to increase.

The number of formula errors that are possible in R is directly related to the total number of formula parameters required to correctly write each formula. In a spreadsheet, the number of formula errors is a function of both the number of formula parameters and the number of cell locations needed to represent the full response range of results. Can we make errors in R-based analysis? Of course, but the potential for those errors is exponentially larger in spreadsheets.

As we've already seen, too, R operates according to a linear flow that guides the development of logic. Also, variables can be named in a way that makes sense to the context of the problem[5] so that the program formulation and business logic are more closely merged, reducing the burden of inference about the meaning of formulas for auditors and other users. In Chapter 2, I'll present a style guide that will help you maintain clarity in the definition of variables, functions, and files.

Although R answers the concerns of direct cost and the propagation of formula errors, its procedural language structure presents a higher barrier to improper use because it requires a more rational, structured logic than is required by spreadsheets, requiring a rigor that people usually learn from programming and software design. The best aspect of R is that it communicates the formulation and logic layer of an analysis in a more straightforward manner as the procedural instructions for performing calculations. It preserves the flow of thought that is necessary to move from starting assumptions to conclusions. The numerical layer is presented only when requested, but logic and formulation are more visibly available. As we move forward through this tutorial, I'll explain more about how these features present themselves for effective business case analysis.

---

[5]Spreadsheets allow the use of named references, but the naming convention can become unwieldy if sections in an array need different names.

What You Will Learn

Part 1 of this book presents a tutorial for learning how to use the statistical programming language R to develop a business case simulation and analysis. I assume you possess at least the skill level of a novice R user.

The tutorial addresses the case in which a chemical manufacturing company considers constructing a new chemical reactor and production facility to bring a new compound to market. There are several uncertainties and risks involved, including the possibility that a competitor will bring a similar product online. The company must determine the value of making the decision to move forward as well as where they might prioritize their attention to make a more informed and robust decision.

The purpose of the book is not to teach you R in a broad manner. There are plenty of resources that do that well now. Rather, it attempts to show you how to do the following:

- Set up a business case abstraction for clear communication of the analysis.

- Model the inherent uncertainties and resultant risks in the problem with Monte Carlo simulation.

- Communicate the results graphically.

- Draw appropriate insights from the results.

So, although you will not necessarily become an R power user, you will gain some insights into how to use this powerful language to escape the foolish consistency of spreadsheet dependency. There is a better way.

What You Will Need

To follow this tutorial, you will need to download and install the latest version of R for your particular operating system. R can be obtained here at `http://www.r-project.org/`. Because I wrote this tutorial with the near beginner in mind, you will only need the base installation of R and no additional packages. Having said that, I encourage you to learn to use the integrated development environment (IDE) RStudio with many of the third-party packages that are available for the R language as quickly as possible to take advantage of workflow, presentation, and processing improvements over the base R installation. For now, though, let's focus on getting the key concepts without the burden of simultaneously learning the feature nuances of a peculiar third-party environment.

CHAPTER 2

Setting Up the Analysis

The purpose of analysis is to produce and communicate effectively and clearly helpful insights about a problem. Good analysis begins with good architecture and good housekeeping of the analysis structure, elements, relationships, and style.

The Case Study

The following is a simple case study that I often use when I teach seminars on quantitative business analysis. It's simple enough to convey a big-picture idea, yet challenging enough to teach important concepts related to structuring the thoughts and communications about complex business interactions, representing uncertainty, and managing risk. There is no doubt the business case could contain many more important details than are presented here. The idea, however, is not to develop an exhaustive template to use in every case; rather, the key idea is to demonstrate the use of the R language in just enough detail to build out your own unique cases as needed.

We will think through the case study in two sections. The first section, "Deterministic Base Case," describes the basic problem for the business analysis with single value assumptions. These assumptions are used to set up the skeletal framework of the model. Once we set up and validate this deterministic framework, we then expand the analysis to include the consideration of uncertainties presented in the second section, "The Risk Layer," that might expose our business concept to undesirable outcomes.

Deterministic Base Case

ACME-Chem Company is considering the development of a chemical reactor and production plant to deliver a new plastic compound to the market. The marketing department estimates that the market can absorb a total of 5 kilotons a year when it is mature, but it will take five years to reach that maturity from halfway through

© Robert D. Brown III 2018
R. D. Brown III, *Business Case Analysis with R*, https://doi.org/10.1007/978-1-4842-3495-2_2

construction, which occurs in two phases. They estimate that the market will bear a price of $6 per pound.

Capital spending on the project will be $20 million per year for each year of the first phase of development and $10 million per year for each year of the second phase. Management estimates that development will last four years, two years for each phase. After that, a maintenance capital spending rate will be $2 million per year over the life of the operations. Assume a seven-year straight-line depreciation schedule for these capital expenditures. (To be honest, a seven-year schedule is too short in reality for these kinds of expenditures, but it will help illustrate the results of the depreciation routine better than a longer schedule.)

Production costs are estimated to be fixed at $3 million per year, but will escalate at 3% annually. Variable component costs will be $3.50 per pound, but cost reductions are estimated to be 5% annually. General, sales, and administrative (GS&A) overhead will be around 20% of sales.

The tax rate is 38%. The cost of capital is 12%. The analytic horizon is 20 years. Determine the following:

1. Cash flow profile.

2. Cumulative cash flow profile.

3. Net present value (NPV) of the cash flow.

4. Pro forma table.

5. Sensitivity of the NPV to low and high changes in assumptions.

The Risk Layer

The market intelligence group has just learned that RoadRunner Ltd. is developing a competitive product. Marketing believes that there is a 60% chance RoadRunner will also launch a production facility in the next four to six years. If they get to market before ACME-Chem, half the market share will be available to ACME-Chem. If they are later, ACME-Chem will maintain 75% of the market. In either case, the price pressure will reduce the monopoly market price by 15%.

What other assumptions should be treated as uncertainties in the base case?

Show the following:

1. Cash flow and cumulative cash flow with confidence bands.

2. Histogram of NPV.

3. Cumulative probability distribution of NPV.

4. Waterfall chart of the pro forma table line items.

5. Tornado sensitivity chart of 80th percentile ranges in uncertainties.

Abstract the Case Study with an Influence Diagram

When I ask people what the goal of business analysis should be, they usually respond with something like logical consistency or correspondence to the real world. Of course, I don't disagree that those descriptors are required features of good business analysis; however, no one ever commissions business case analysis to satisfy intellectual curiosity alone. Few business decision makers ever marvel over the intricate nature of code and mathematical logic. They do, however, marvel at the ability to produce clarity while everyone else faces ambiguity and clouded thinking.

I assert that the goal of business case analysis is to produce and effectively communicate clear insights to support decision making in a business context. Clear communication about the context of the problem at hand and how insights are analytically derived is as important, if not more so, than logical consistency and correspondence. Although it is certainly true that clear communication is not possible without logical consistency and correspondence, logical consistency and correspondence are almost useless unless their implications are clearly communicated.

I think the reason many people forget this issue of clear communication is that, as analysts who love to do analysis, we tend to assume lazily that our analysis speaks for itself in the language we are accustomed to using among each other. We tend to forget that the output of our thinking is a product to be used by others at a layer that does not include the underlying technical mechanisms. Imagine being given an iPhone that requires the R&D laboratory to operate. An extremely small number of users would ever be able to employ such a device. Instead, Apple configures the final product to present a much simpler user interface to the target consumers than design engineers employ, yet they still have access to the power that the underlying complexity supports. Likewise, good business case analysis–that which supports the clear communication of insights to support decision making in a business context–should follow principles of good product design.

Before we write the first line of R code (or any code in any construct, even a spreadsheet) I recommend that we translate the context of the problem we have been asked to analyze to a type of flowchart that communicates the essence of the problem. This flowchart is called an *influence diagram.*

Influence diagrams are not procedural flowcharts. They do not tell someone the order of operations that would be implemented in code. They do, however, graphically reveal how information (both fixed and uncertain) and decisions coordinate through intermediate effects to affect the value of some target key figures of merit on which informed decisions are based. In most business case analyses, the target key figures of merit (or objective functions) tend to be a cash flow profile and its corresponding NPV.

The influence diagram shown in Figure 2-1 captures the essence of the deterministic base case. It reveals how starting assumptions work through interdependent calculations to the objective function, NPV. The structure of the influence diagram is built by connecting the assumptions to other nodes with solid arcs to indicate the flow of causality or conditionality. Intermediate nodes are then connected to other dependent nodes in sequence until some set of nodes converge on the objective or value node.

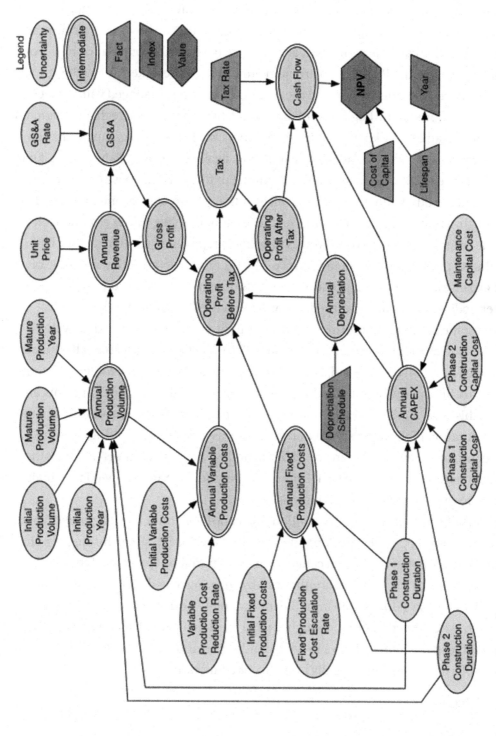

Figure 2-1. *The influence diagram of the deterministic base case*

Notice that assumptions are represented both as light blue ellipses and purple trapezoids. The light blue ellipses represent assumptions that are not necessarily innately fixed in value. They can change in the real world due to effects beyond our control. For example, notice that the Phase 1 construction duration is represented as an ellipse. Although it is true that this outcome in the real world might be managed with project management efforts, at the beginning of the project, no one can declare that the outcome will conform to an exactly known value. Many events, such as weather, supply chain failures, mechanical failures, and so on, could work against the desired duration. Likewise, some events could influence a favorable duration. For this reason, Phase 1 construction duration and the other light blue ellipses will eventually be treated as uncertainties. For now, though, as we set up the problem, we treat their underlying values as tentatively fixed. The other assumptions displayed as trapezoids represent facts that are fixed by constraints of the situation, by definition (i.e., the tax rate or depreciation schedule of capital expense), or are known physical constants or unit conversion constants.

Intermediate values are displayed as double-lined yellow ellipses. These nodes represent some operation performed on assumptions and other intermediate values along the value chain to the objective function. The objective function (or value function), usually a red hexagon or diamond, is the ultimate target of all the preceding operations.

The dark blue parallelogram in the influence diagram labeled Year represents an index. In this case, it is an index for a time axis defined by the initial assumption of the life span duration of the analysis but that functions as a basis for all the other values. Because this node can affect many or all of the nodes, we don't connect it to the other dependent nodes to reduce the visual complexity of the diagram. We need to explain this condition to the consumers of our analysis.

Historically, we observe the convention that canonical influence diagrams do not contain feedback loops or cyclic dependencies. As much as possible, the diagram should represent a snapshot of the flow of causality or conditionality. However, if the clarity of the abstraction of the problem is enhanced by showing a feedback loop in a time domain, we can use a dotted line from the intermediate node back to an appropriate predecessor node. Keep in mind, though, that in the actual construction of your code, some variable in the loop will have to be dependent on another variable with a lagged subscript of a form that conceptually looks like the function

```
Var2 = f(Var1[t-n])
```

where t is the temporal index, and n is the lag constant.

Again, the influence diagram is supposed to capture the essence of the problem in an abstract form that does not reveal every detail of the procedural line code you will employ. It simply gives the big picture. Constructed carefully, though, the diagram will provide you as the analyst an inventory of the classes of key variables required in your code. You should also make a habit of defining these variables in a table that includes their units and subject matter experts (SMEs) or other resources consulted that inform the definitions or values used.

The influence diagram displayed in Figure 2-1 represents the base case deterministic model; that is, the model that does not include the risk considerations. Once we develop the base case code structure and produce our preliminary analysis, we will modify the influence diagram to include those elements that capture our hypotheses about the sources of risk in the problem.

Set Up the File Structure

Again, we should think of good business case analysis as a product for some kind of customer. As much as possible, all aspects of the development of our analysis should seek to satisfy the customer's needs in regard to the analysis. Usually, the customers we have in mind are the decision makers who will use our analysis to make commitments or allocate resources. Decision makers might not be the only consumers of our analysis, though. There are a number of reasons why other analysts might use our analysis as well, but if we are not immediately or no longer available to clarify ambiguous code, their job becomes that much more difficult. There is a social cost that accompanies ambiguity, and my personal ethics guide me to minimize those costs as much as possible. With that in mind, I recommend setting up a file structure for your analysis code that will provide guidance about the content of various files.

The file directory that contains your analysis project should be appropriately and clearly named (e.g., BizSimWithR). Within that directory, create at least three other subdirectories:

1. *Data:* Contains text and comma-separated value (CSV) files that store key assumptions for your analysis.

2. *Libraries:* Contains text files of functions and other classes you might import that support the operation of your code.

3. *Results:* Contains output files of results you export from within your analysis (maybe for postprocessing with other tools) as well as graphic files that might be used in reporting the results of your analysis.

To avoid the problem of resetting the working directory of your projects, I also recommend that you build your file structure in the R application working directory. You can do this manually the way you would create any other file directory or you can use the `dir.create()` function of R. Consult the R Help function for the details of `dir.create()`, but to quickly and safely set up your project file directory, use the following commands in the R console:

```
dir.create("BizSimWithR", recursive = F)
dir.create("BizSimWithR/data", recursive = F)
dir.create("BizSimWithR/libraries", recursive = F)
dir.create("BizSimWithR/results", recursive = F)
```

Of course, you should create any other subdirectories that are needed to partition categories of files for good housekeeping.

Style Guide

Keeping with the theme of clear communication out of consideration for those who consume your code directly, I also recommend that you follow a style guide for syntax, file, variable, and function names within your R project. Google's R Style Guide[1] provides an excellent basis for naming conventions. Following this guide provides one more set of clues to the nature of the code others might need to comprehend.

In short, the Google convention follows these guidelines:

1. Separate words in the name of the file with an underscore.

2. End file names with `.R`.

3. Name variable identifiers.

    Separate words in variables with a period, as in `cash.flow`. (This convention generally conflicts with that of Java and Python variable naming; consequently, its practice is forbidden in some coding shops. If you think your code might be reused in those environments, stick with the most universally applicable naming convention to avoid later frustrations and heartbreaks.)

---

[1]`https://google.github.io/styleguide/Rguide.xml`

- Start each word in a function name with a capital letter. Make the function name a verb, as in `CalculateNPV`.

- Name constants like the words in functions, but append a lowercase k at the beginning, as in `kTaxRate`.

4. Use the following syntax.

    - The maximum line length should be 80 characters.

    - Use two spaces to indent code. (Admittedly, I violate this one, as I like to use the tab to indent. Some rules are just meant to be broken. Most IDEs will now let you define a tab as two spaces in the Preferences pane.)

    - Place a space around operators.

    - Do not place a space before a comma, but always place one after a comma.

    - Do not place an opening curly brace on its own line. Do place a closing curly brace on its own line.

    - Use <-, not =, for variable assignment. (Arguments about the choice of assignment operator in R can approach religious levels of zeal. However, the <- operator is still the most widely adopted convention in R programming circles.)

    - Do not use semicolons to terminate lines or to put multiple commands on the same line.

Of course, always provide clear comments within your code to describe routines for which operation might not be immediately obvious (and never assume they are immediately obvious) and to tie the flow of the code back to the influence diagram. The complete Google R Style Guide presents more complete details and examples as well as other recommendations for file structuring.

I learned the value of these guidelines the hard way several years ago while working on a very complex financial analysis for a new petroleum production field. The work was fast paced, and my client was making frequent change requests. In an effort to save time (and not possessing the maturity to regulate change orders effectively with a client), I quit following a convention for clear communication, namely adding thorough comments to my code. As I was certainly tempting the Fates, I got hit by the bus of gall bladder disease and exited the remainder of the project for surgery and extended recuperation. Although my files were turned over to a very capable colleague, he could make little sense of many important sections of my code. Consequently, our deliverables fell behind schedule, costing us both respect and some money. I really hated losing the respect, even more than the money.

Write the Deterministic Financial Model

We need to write our code in the order of events implied by our influence diagram. Studying the diagram, we can see that the nodes directly related to the annual capital expense (CAPEX) are involved in driving practically everything else. This makes sense because no production occurs until the end of the first phase of construction. No revenues or operational expenses occur, either, until production occurs. Therefore, the first full code block we address will be the CAPEX block. Just before we get to that, though, we need to set up a data file to hold the key assumptions that will be used in the model.

Data File

To begin the coding, we create a data file named, say, `global_assumptions.R` and place it in the `~/BizSimWithR/data` subdirectory of our coding project. The contents of the data file look like this:

```
# Global Model Assumptions
kHorizon <- 20 # years, length of time the model covers.
year <- 1:kHorizon # an index for temporal calculations.
kSampsize <- 1000 # the number of iterations in the Monte Carlo simulation.
run <- 1:kSampsize # the iteration index.
kTaxRate <- 38 # %
kDiscountRate <- 12 # %/year, used for discounted cash flow calculations.
kDeprPer <- 7 # years, the depreciation schedule for the capital.
```

The values in this file are those that we will treat as constants, constraints, or assumptions that are used throughout the model. Notice that next to each assumption we comment on the units employed and add a brief description. Also, we choose variable identifiers that are descriptive and will permit easy interpretation by another user who might need to pick up the analysis after becoming familiar with the context of the problem.

Next, we import this file into our main R text file (`Determ_Model.R`) using the `source()` command.

```
source("~/BizSimWithR/data/global_assumptions.R")
```

If you are familiar with the concept of the server-side include in HTML programming, you will easily understand the `source()` command. It essentially reads the contents as written in the file into the R session for processing.

For the variables that we will perform sensitivity analysis around or ultimately treat as uncertainties, we create a CSV data file named `risk_assumptions.csv` that contains the values of the operational variables used in the deterministic and risk model (Figure 2-2).

◇	A	B	C	D	E	F
1	variable	p10	p50	p90	units	notes
2	p1.capex	35000000	40000000	47000000	$	phase 1 capital spend to be spread over the duration of Phase 1.
3	p1.dur	1	2	4	yrs	phase 1 duration.
4	p2.capex	15000000	20000000	30000000	$	phase 2 capital spend to be spread over the duration of Phase 2.
5	p2.dur	1	2	4	yrs	phase 2 duration.
6	maint.capex	1500000	2000000	3000000	$	annual maintenance spend.
7	fixed.prod.cost	2250000	3000000	4500000	$	annual fixed production costs.
8	prod.cost.escal	2.25	3	4.5	%/yr	the rate at which the fixed costs grow.
9	var.prod.cost	2.625	3.5	5.25	$/lb	the variable production costs.
10	var.cost.redux	3.75	5	7.5	%/yr	the rate at which the variable production costs decay.
11	gsa.rate	19	20	22	% revenue	general, sales & administration cost.
12	time.to.peak.sales	3.75	5	7.5	yrs	the number of years to reach max sales.
13	mkt.demand	3.75	5	7.5	ktons/yr	the amount of annual sales at peak.
14	price	4.5	6	9	$/lb	product price.
15	rr.comes.to.market	0	0.6	1		the probability that RoadRunner comes to market.
16	rr.time.to.market	4	5	6	yrs	the time elapsed until RoadRunner comes to market.
17	early.market.share	45	50	65	%	the market share available to Acme if RoadRunner comes to market early.
18	late.market.share	70	75	85	%	the market share available to Acme if RoadRunner comes to market late.
19	price.redux	12	15	20	%	reduction in price that that occurs if RoadRunner comes to market.

Figure 2-2. *The data table used for this business case in* `.csv` *format rendered in a spreadsheet*

We create columns in this file not only to capture the names of the variables and their values, but their units and explanatory notes, as well.

We read the contents of the CSV file using the `read.csv()` command and assign it to a variable.

```
d.data <- read.csv("~/BizSimWithR/data/risk_assumptions.csv")
```

The `read.csv()` function reads the data in the table structure of the file, then creates a data frame from it such that the header row names the columns, and the rows are created from the case values in the table. Anytime we need to refer to the contents of the assumption file while in the R console, we can do so quickly by calling `d.data`.

```
> d.data
              variable       p10      p50      p90     units                                                         notes
1           p1.capex 3.500e+07 4.0e+07 4.70e+07        $    phase 1 capital spend to be spread over the duration of Phase 1.
2             p1.dur 1.000e+00 2.0e+00 4.00e+00      yrs                                                phase 1 duration.
3           p2.capex 1.500e+07 2.0e+07 3.00e+07        $    phase 2 capital spend to be spread over the duration of Phase 1.
4             p2.dur 1.000e+00 2.0e+00 4.00e+00      yrs                                                phase 2 duration.
5         maint.capex 1.500e+06 2.0e+06 3.00e+06        $                                       annual maintenance spend.
6      fixed.prod.cost 2.250e+06 3.0e+06 4.50e+06        $                                  annual fixed production costs.
7       prod.cost.escal 2.250e+00 3.0e+00 4.50e+00    %/yr                             the rate at which the fixed costs grow.
8        var.prod.cost 2.625e+00 3.5e+00 5.25e+00    $/lb                                   the variable production costs.
9       var.cost.redux 3.750e+00 5.0e+00 7.50e+00    %/yr         the rate at which the variable production costs decay.
10            gsa.rate 1.900e+01 2.0e+01 2.20e+01 % revenue                          general, sales & administration cost.
11  time.to.peak.sales 3.750e+00 5.0e+00 7.50e+00      yrs                           the number of years to reach max sales.
12          mkt.demand 3.750e+00 5.0e+00 7.50e+00 ktons/yr                            the amount of annual sales at peak.
13               price 4.500e+00 6.0e+00 9.00e+00    $/lb                                               product price.
14   rr.comes.to.market 0.000e+00 6.0e-01 1.00e+00                   the probability that RoadRunner comes to market.
15   rr.time.to.market 4.000e+00 5.0e+00 6.00e+00      yrs               the time elapsed until RoadRunner comes to market.
16  early.market.share 4.500e+01 5.0e+01 6.50e+01        %  the market share available to Acme if RoadRunner comes to market early.
17   late.market.share 7.000e+01 7.5e+01 8.50e+01        %   the market share available to Acme if RoadRunner comes to market late.
18          price.redux 1.200e+01 1.5e+01 2.00e+01        %        reduction in price that that occurs if RoadRunner comes to market.
```

The data frame looks pretty much the same as the CSV file structure as it appears in Microsoft Excel; however, we will need to use the p50 values only for the initial deterministic analysis, so we slice out the p50 values from `d.data` and place them in their own vector. For the deterministic analysis, we will use only values of the first 13 variables. We will use the last five variables in the risk layer of the model.

```
# Slice the p50 values from data frame d.data.
d.vals <- d.data$p50[1:13]
```

The contents of the p50 vector look like this:

```
[1] 4.0e+07 2.0e+00 2.0e+07 2.0e+00 2.0e+06 3.0e+06 3.0e+00 3.5e+00 5.0e+00
2.0e+01 5.0e+00 5.0e+00 6.0e+00
```

The final step in the process of setting up the data for our deterministic analysis is to assign the p50 values in `d.vals` to variable names in the R file.

```
# Assign p50 values to variables.
p1.capex <- d.vals[1]
p1.dur <- d.vals[2]
p2.capex <- d.vals[3]
p2.dur <- d.vals[4]
maint.capex <- d.vals[5]
```

```
fixed.prod.cost <- d.vals[6]
prod.cost.escal <- d.vals[7]
var.prod.cost <- d.vals[8]
var.cost.redux <- d.vals[9]
gsa.rate <- d.vals[10]
time.to.peak.sales <- d.vals[11]
mkt.demand <- d.vals[12]
price <- d.vals[13]
```

At this point, you might be wondering what the p10, p50, p90 values represent, as the deterministic base case description of the model only included single point values. Remember, though, that in the description of the influence diagram, I described the ellipses as representatives of uncertain variables. For the deterministic base case, we treat these values in a fixed manner. Ultimately, however, for the risk layer of the business case to be thoroughly considered, ranges for each of the variables will be assessed by SMEs as their 80th percentile prediction interval. The three characteristic points of those prediction intervals are the 10th, 50th, and 90th percentiles. We use the median values, the p50s, for the deterministic analysis. The p50s were the numbers used in the business case descriptions.

CAPEX Block

The first thing we need to establish in the actual calculation of the model is when the capital gets expended by phase. The phase calculation accomplishes this by telling R when the construction and operating phases occur. We can use 1 and 2 for the construction phases, and 3 for the maintenance phase.

```
# CAPEX Module
phase <- (year <= p1.dur) * 1 +
    (year > p1.dur & year <= (p1.dur + p2.dur)) * 2 +
    (year > (p1.dur + p2.dur)) * 3
```

The result is a vector across the year index.

```
[1] 1 1 2 2 3 3 3 3 3 3 3 3 3 3 3 3 3 3 3 3
```

From our influence diagram, we know that capex is a vector conditionally related to phase, captured by the following expression.

```
capex <- (phase == 1) * p1.capex / p1.dur +
    (phase == 2) * p2.capex / p2.dur +
    (phase == 3) * maint.capex
```

It produces the following vector.

```
[1] 2e+07 2e+07 1e+07 1e+07 2e+06 2e+06 2e+06 2e+06 2e+06 2e+06 2e+06 2e+06
2e+06 2e+06 2e+06 2e+06 2e+06 2e+06 2e+06 2e+06
```

Next, we need to handle the depreciation that will be subtracted from the gross profit to find our taxable income. The depreciation is based on the capital emplaced at the time of construction or when it is incurred (i.e., maintenance); however, it is not taken into account until it can be applied against taxable profit. This means that the capital incurred in Phase 1 won't be amortized until Phase 1 is over and the plant begins generating revenue. Capital incurred in the Phase 2 construction and maintenance phases can be amortized starting in the year following each year of expenditure.

One way to handle this operation would be to use a for loop. The following code performs this process as already described.

We start with

```
depr.matrix <- matrix(rep(year, kHorizon), nrow = kHorizon,
    ncol = kHorizon, byrow = TRUE)
```

to initialize the following matrix.

```
        [,1] [,2] [,3] [,4] [,5] [,6] [,7] [,8] [,9] [,10] [,11] [,12] [,13] [,14] [,15] [,16] [,17] [,18] [,19] [,20]
 [1,]     1    2    3    4    5    6    7    8    9    10    11    12    13    14    15    16    17    18    19    20
 [2,]     1    2    3    4    5    6    7    8    9    10    11    12    13    14    15    16    17    18    19    20
 [3,]     1    2    3    4    5    6    7    8    9    10    11    12    13    14    15    16    17    18    19    20
 [4,]     1    2    3    4    5    6    7    8    9    10    11    12    13    14    15    16    17    18    19    20
 [5,]     1    2    3    4    5    6    7    8    9    10    11    12    13    14    15    16    17    18    19    20
 [6,]     1    2    3    4    5    6    7    8    9    10    11    12    13    14    15    16    17    18    19    20
 [7,]     1    2    3    4    5    6    7    8    9    10    11    12    13    14    15    16    17    18    19    20
 [8,]     1    2    3    4    5    6    7    8    9    10    11    12    13    14    15    16    17    18    19    20
 [9,]     1    2    3    4    5    6    7    8    9    10    11    12    13    14    15    16    17    18    19    20
[10,]     1    2    3    4    5    6    7    8    9    10    11    12    13    14    15    16    17    18    19    20
[11,]     1    2    3    4    5    6    7    8    9    10    11    12    13    14    15    16    17    18    19    20
[12,]     1    2    3    4    5    6    7    8    9    10    11    12    13    14    15    16    17    18    19    20
[13,]     1    2    3    4    5    6    7    8    9    10    11    12    13    14    15    16    17    18    19    20
[14,]     1    2    3    4    5    6    7    8    9    10    11    12    13    14    15    16    17    18    19    20
[15,]     1    2    3    4    5    6    7    8    9    10    11    12    13    14    15    16    17    18    19    20
[16,]     1    2    3    4    5    6    7    8    9    10    11    12    13    14    15    16    17    18    19    20
[17,]     1    2    3    4    5    6    7    8    9    10    11    12    13    14    15    16    17    18    19    20
[18,]     1    2    3    4    5    6    7    8    9    10    11    12    13    14    15    16    17    18    19    20
[19,]     1    2    3    4    5    6    7    8    9    10    11    12    13    14    15    16    17    18    19    20
[20,]     1    2    3    4    5    6    7    8    9    10    11    12    13    14    15    16    17    18    19    20
```

Then, we run a loop to set up a depreciation schedule programmatically (instead of typing each one manually):

```
for (y in year) {
    if (y <= p1.dur) {
        depr.matrix[y, ] <- 0
    } else if (y == (p1.dur+1)) {
        depr.matrix[y, ] <- (depr.matrix[y, ] >= (1 + p1.dur)) *
            (depr.matrix[y, ] < (y+ kDeprPer)) * p1.capex / kDeprPer
    } else {
        depr.matrix[y, ] <- (depr.matrix[y, ] >= y) *
            (depr.matrix[y, ] < (y + kDeprPer)) * capex[y - 1] / kDeprPer
    }
}
```

This loop incorporates the logic that, in general, for a straight-line depreciation schedule of n years, capital C incurred in year Y is spread over years Y + 1, Y + 2, …, Y + n - 1, as C/n in each year. So, if the project incurs $10 million of construction capital in Year 3, a seven-year depreciation schedule would spread that out evenly over seven years ($10 million/7 years) starting in Year 4 and running through Year 10.

More specifically to our earlier depreciation code snippet, R looks in the initialization table and replaces the numbers in rows for the Phase 1 duration years with 0s. Starting in the row representing the first year after the end of Phase 1 when sales are produced and a taxable profit can be made, our code replaces any value less than or equal to the Phase 1 duration with a 0. Then it looks for values in the row that are greater than the Phase 1 duration and less than or equal to the Phase 1 duration plus the depreciation period, and it replaces those values with the appropriate capital value divided by the depreciation period. Finally, the remainder of values in the row are set to 0. This process is repeated for each phase. The result of our code block for depreciation produces the matrix shown in Figure 2-3.

```
> depr.matrix
       [,1] [,2]    [,3]    [,4]    [,5]      [,6]      [,7]      [,8]      [,9]     [,10]
 [1,]    0    0       0       0       0       0.0       0.0       0.0       0.0       0.0
 [2,]    0    0       0       0       0       0.0       0.0       0.0       0.0       0.0
 [3,]    0    0 5714286 5714286 5714286 5714285.7 5714285.7 5714285.7 5714285.7       0.0
 [4,]    0    0       0 1428571 1428571 1428571.4 1428571.4 1428571.4 1428571.4 1428571.4
 [5,]    0    0       0       0 1428571 1428571.4 1428571.4 1428571.4 1428571.4 1428571.4
 [6,]    0    0       0       0       0  285714.3  285714.3  285714.3  285714.3  285714.3
 [7,]    0    0       0       0       0       0.0  285714.3  285714.3  285714.3  285714.3
 [8,]    0    0       0       0       0       0.0       0.0  285714.3  285714.3  285714.3
 [9,]    0    0       0       0       0       0.0       0.0       0.0  285714.3  285714.3
[10,]    0    0       0       0       0       0.0       0.0       0.0       0.0  285714.3
[11,]    0    0       0       0       0       0.0       0.0       0.0       0.0       0.0
[12,]    0    0       0       0       0       0.0       0.0       0.0       0.0       0.0
[13,]    0    0       0       0       0       0.0       0.0       0.0       0.0       0.0
[14,]    0    0       0       0       0       0.0       0.0       0.0       0.0       0.0
[15,]    0    0       0       0       0       0.0       0.0       0.0       0.0       0.0
[16,]    0    0       0       0       0       0.0       0.0       0.0       0.0       0.0
[17,]    0    0       0       0       0       0.0       0.0       0.0       0.0       0.0
[18,]    0    0       0       0       0       0.0       0.0       0.0       0.0       0.0
[19,]    0    0       0       0       0       0.0       0.0       0.0       0.0       0.0
[20,]    0    0       0       0       0       0.0       0.0       0.0       0.0       0.0


      [,11]     [,12]     [,13]     [,14]     [,15]     [,16]     [,17]     [,18]     [,19]     [,20]
        0.0       0.0       0.0       0.0       0.0       0.0       0.0       0.0       0.0       0.0
        0.0       0.0       0.0       0.0       0.0       0.0       0.0       0.0       0.0       0.0
        0.0       0.0       0.0       0.0       0.0       0.0       0.0       0.0       0.0       0.0
        0.0       0.0       0.0       0.0       0.0       0.0       0.0       0.0       0.0       0.0
  1428571.4       0.0       0.0       0.0       0.0       0.0       0.0       0.0       0.0       0.0
   285714.3  285714.3       0.0       0.0       0.0       0.0       0.0       0.0       0.0       0.0
   285714.3  285714.3  285714.3       0.0       0.0       0.0       0.0       0.0       0.0       0.0
   285714.3  285714.3  285714.3  285714.3       0.0       0.0       0.0       0.0       0.0       0.0
   285714.3  285714.3  285714.3  285714.3  285714.3       0.0       0.0       0.0       0.0       0.0
   285714.3  285714.3  285714.3  285714.3  285714.3  285714.3       0.0       0.0       0.0       0.0
   285714.3  285714.3  285714.3  285714.3  285714.3  285714.3  285714.3       0.0       0.0       0.0
        0.0  285714.3  285714.3  285714.3  285714.3  285714.3  285714.3  285714.3       0.0       0.0
        0.0       0.0  285714.3  285714.3  285714.3  285714.3  285714.3  285714.3  285714.3       0.0
        0.0       0.0       0.0  285714.3  285714.3  285714.3  285714.3  285714.3  285714.3  285714.3
        0.0       0.0       0.0       0.0  285714.3  285714.3  285714.3  285714.3  285714.3  285714.3
        0.0       0.0       0.0       0.0       0.0  285714.3  285714.3  285714.3  285714.3  285714.3
        0.0       0.0       0.0       0.0       0.0       0.0  285714.3  285714.3  285714.3  285714.3
        0.0       0.0       0.0       0.0       0.0       0.0       0.0  285714.3  285714.3  285714.3
        0.0       0.0       0.0       0.0       0.0       0.0       0.0       0.0  285714.3  285714.3
        0.0       0.0       0.0       0.0       0.0       0.0       0.0       0.0       0.0  285714.3
```

Figure 2-3. *The depreciation table for the capital costs over the model time horizon*

Now, if we sum the columns across the years of depr.matrix, we get the total annual depreciation.

```
# Sum the columns of the depreciation matrix to find the annual
# depreciation.
depr <- colSums(depr.matrix)
```

Because R is a vectorized language, though, many people frown on the use of for loops. Instead we might consider replacing the loop with a variant of the apply() function combined with the ifelse() functions. Here, we use the sapply() function. The sapply() function is a type of functional for loop that steps across the elements of an index, applying the value of the step at the appropriate place in the expression defined after the function() statement.

A general for loop does this:

```
for (i in index) {
    expr[..., i, ...]
}
```

The sapply() function does this instead:

```
sapply(index, function(i) expr[..., i, ...]).
```

It's a great way to reduce the visual complexity of the expressions we write as well as accelerate the speed of the code. We use this function more in the sections to come. In the meantime, to learn more about sapply(), simply call the help function in the R console using ?sapply or help(sapply).

Now, we replace the depreciation code block that employs the for loop with this sapply() function and nested ifelse() functions to accommodate the conditional requirements in the loop. Note that we transpose the results of sapply() with t() because the sapply() function produces a matrix result that is transposed from the for loop approach.

```
# Depreciation Module
depr.matrix <-
  t(sapply(year, function(y)
    ifelse(
      y <= p1.dur & year > 0,
      0,
```

```
   ifelse(
     y == (p1.dur + 1) & year < y + kDeprPer & year >= y,
     p1.capex / kDeprPer,
     ifelse((year >= y) & (year < (y + kDeprPer)),
             capex[y - 1] / kDeprPer, 0)
   )
 )
  )
)
depr <- colSums(depr.matrix)
```

We can now see what the depreciation calculation looks like graphically (Figure 2-4) with the plot() function using

```
plot(year,
      depr / 1e6,
      xlab = "Year",
      ylab = "Depreciation [$000,000]",
      type = "b")
```

Note that we scaled the depr values by 1 million to make the plot a little more readable, using the "b" parameter to plot *both* lines and points.

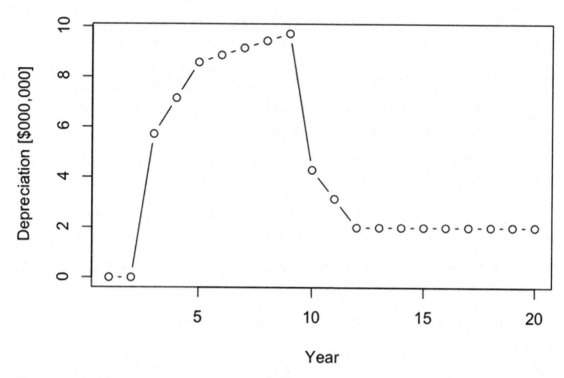

Figure 2-4. *The total depreciation over time*

Sales and Revenue Block

Sales begin in the year following the end of construction in Phase 1. Here we model the market adoption as a straight line over the duration from the start of Phase 2 to the time of peak sales.[2]

```
mkt.adoption <- pmin(cumsum(phase > 1) / time.to.peak.sales, 1).
```

(phase > 1) produces a vector of TRUEs starting from the position in the phase vector where Phase 2 starts, and FALSEs prior to that.

---

[2]A straight line is likely not the best model for this. My inclination would be to use an absorption model in which sales start slowly during the market introduction phase, ramp up quickly at some point as the market matures, then level off as the remaining amounts of demand are realized. This approach is easily accomplished with the following function: SCurv = function(y0, time, time.to.peak) {this.s.curve = 1 / (1 + (y0 / (1 - y0)) ^ (2 * time/time.to.peak - 1)) }, where y0 is the initial absorption as a fraction of the total, time is an index across which the absorption responds, and time.to.peak is the amount of time it takes to go from y0 to 1 - y0.

The cumsum(phase > 1) phrase coerces the logical states to integers (1 for TRUE, and 0 for FALSE), then cumulates the vector to produce a result that looks like this:

```
[1]  0  0  1  2  3  4  5  6  7  8  9  10  11  12  13  14  15  16  17  18.
```

By dividing this vector by time.to.peak.sales, we get a vector that normalizes this cumulative sum to the duration of time required to reach peak sales.

```
[1] 0.0 0.0 0.2 0.4 0.6 0.8 1.0 1.2 1.4 1.6 1.8 2.0 2.2 2.4 2.6 2.8 3.0 3.2
3.4 3.6
```

The pmin() function performs a pairwise comparison between each number in this vector and the maximum adoption rate of 100%.

```
[1] 0.0 0.0 0.2 0.4 0.6 0.8 1.0 1.0 1.0 1.0 1.0 1.0 1.0 1.0 1.0 1.0 1.0 1.0
1.0 1.0
```

Multiplying this adoption curve by the maximum market demand gives us the sales in each year.

```
sales <- mkt.adoption * mkt.demand * 1000 * 2000
```

Because the price of the product is given in "$/lb," we convert the sales to pounds by multiplying it by the conversion factors 1,000 tons/kiloton and 2,000 lbs/ton.

By plotting revenue with

```
plot(year, sales / 1000,
      xlab = "Year",
      ylab = "Sales [000 lbs]",
      type = "b")
```

we observe the revenue profile (Figure 2-5) and confirm that it is behaving as we expect.

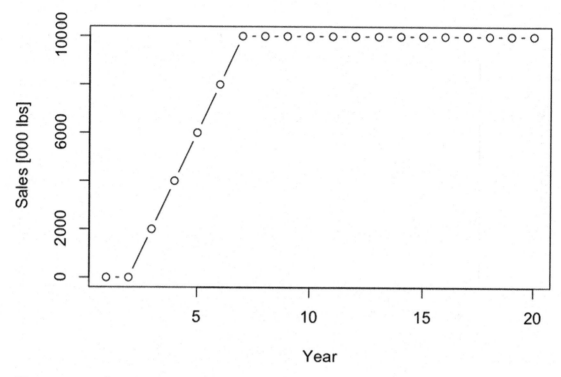

Figure 2-5. *The annual unit sales response*

Finally, revenue is given by

```
revenue <- sales * price.
```

By plotting revenue with

```
plot(year,
     revenue / 1000,
     xlab = "Year",
     ylab = "Revenue [$000]",
     type = "b")
```

we also observe the revenue profile (Figure 2-6) and confirm that it is behaving properly.

29

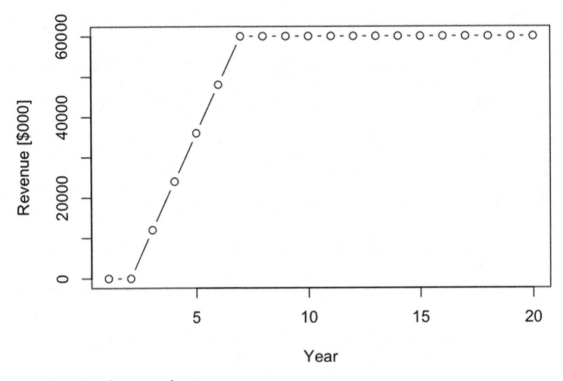

Figure 2-6. *The annual revenue response*

OPEX Block

The operational costs of the business case don't start until after Phase 1. As in the case of the market adoption formula, we establish this with the term (phase > 1) for fixed costs.

```
fixed.cost <- (phase > 1) * fixed.prod.cost * (1 + prod.cost.escal / 100) ^
(year - p1.dur - 1 )
```

We represent the escalation of the fixed costs with the compounding rate term

```
(1 + prod.cost.escal / 100) ^ (year - p1.dur - 1 ).
```

The phrase in the exponent term ensures that the value of the exponent is 0 in Year 3. The combined effect of these two terms produces a series of factors that looks like this:

```
[1] 0.000000 0.000000 1.000000 1.030000 1.060900 1.092727 1.125509 1.159274
1.194052 1.229874 1.266770 1.304773 1.343916 1.384234 1.425761 1.468534
1.512590 1.557967 1.604706 1.652848.
```

Multiplying by the initial fixed cost value, we get

```
[1] 0 0 3000000 3090000 3182700 3278181 3376526 3477822 3582157 3689622
3800310 3914320 4031749 4152702 4277283 4405601 4537769 4673902 4814119
4958543.
```

Because the variable cost is a function of the sales, and we have already defined sales to start after the end of Phase 1, we won't need the (phase > 1) term in the variable cost definition. We do, however, need another compounding factor to account for the decay in the variable cost structure.

```
(1 - var.cost.redux / 100) ^ (year - p1.dur - 1)
```

This series looks like

```
[1] 0 0 1.0000000 0.9500000 0.9025000 0.8573750 0.8145062 0.7737809
0.7350919 0.6983373 0.6634204 0.6302494 0.5987369 0.5688001 0.5403601
0.5133421 0.4876750 0.4632912 0.4401267 0.4181203.
```

The final expression we use for the variable cost takes the form

```
var.cost <- sales * var.prod.cost * (1 - var.cost.redux / 100) ^ (year -
p1.dur - 1)
```

We finish up the remaining expense equations with the following expressions.

```
gsa <- (gsa.rate / 100) * revenue
opex <- fixed.cost + var.cost
```

Pro Forma Block

The pro forma calculations are rather straightforward. According to generally acceptable accounting principles (GAAP)–more or less–we have

- Gross profit = revenue - GS&A

- OPEX = fixed cost + variable cost

- Operating profit before tax = gross profit - OPEX - depreciation

- Operating profit after tax = operating profit before tax - tax

- Cash flow = operating profit after tax + depreciation - CAPEX

Implemented as R code, we have the following:

```
gross.profit <- revenue - gsa
op.profit.before.tax <- gross.profit - opex - depr
tax <- op.profit.before.tax * kTaxRate/100
op.profit.after.tax <- op.profit.before.tax - tax
cash.flow <- op.profit.after.tax + depr - capex
cum.cash.flow <- cumsum(cash.flow)
```

We can see the cash flow (Figure 2-7) in thousands of dollars.

```
plot(year,
     cash.flow / 1000,
     xlab = "Year",
     ylab = "Cash Flow [$000]",
     type = "b")
```

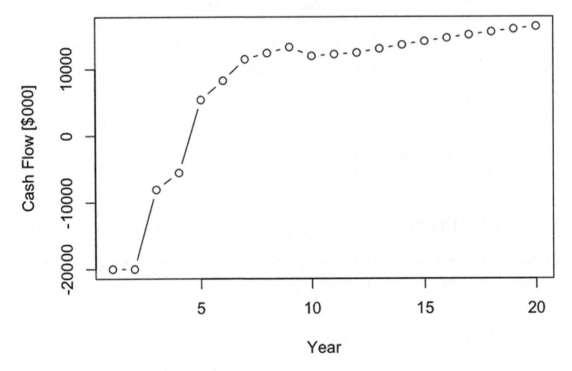

Figure 2-7. *The annual cash flow response*

We can also show the cumulative cash flow (Figure 2-8) in thousands of dollars.

```
plot(year,
     cum.cash.flow / 1000,
     xlab = "Year",
     ylab = "Cash Flow [$000]",
     type = "b")
```

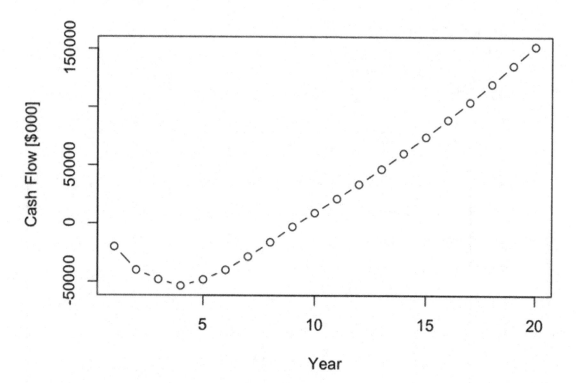

Figure 2-8. *The annual cumulative cash flow*

Net Present Value

To obtain the NPV of the cash flow, we simply need to find the sum of the discounted cash flows over the year index. Following the convention that we count payments as occurring at the end of a time period, we first create a vector of discount factors to be applied to the cash flow.

```
discount.factors <- 1/(1 + kDiscountRate / 100) ^ year
```

which produces

[1] 0.8928571 0.7971939 0.7117802 0.6355181 0.5674269 0.5066311 0.4523492
0.4038832 0.3606100 0.3219732 0.2874761 0.2566751 0.2291742 0.2046198
0.1826963 0.1631217 0.1456443 0.1300396 0.1161068 0.1036668.

Figure 2-9 shows us the plot of the discount.factors.

```
plot(year,
        discount.factors,
        xlab = "Year",
        ylab = "Discount Factors",
        type = "b")
```

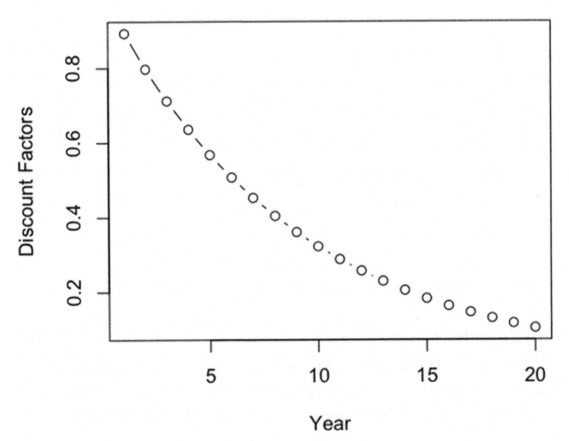

Figure 2-9. *The discount factor profile*

The discounted cash flow is then given by

```
discounted.cash.flow <- cash.flow * discount.factors
```

which produces

```
[1] -17857143 -15943878 -5748744 -3523004 3058044 4183361 5186660 5006359
4788540 3846642 3507746 3187060 2985618 2783313 2583740 2389570 2202730
2024555 1855908 1697283.
```

Figure 2-10 shows us the plot of the `discounted.cash.flow`.

```
plot(year,
     discounted.cash.flow,
     xlab = "Year",
     ylab = "Discounted Cash Flow [$000]",
     type = "b")
```

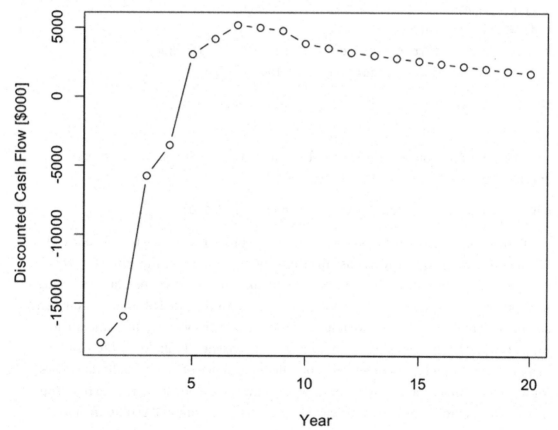

Figure 2-10. *The annual discounted cash flow*

Finally, we sum the values in this vector

```
npv <- sum(discounted.cash.flow)
```

to obtain NPV = $8,214,363.

It is very important to remember at this point that the NPV reported here is based on the deterministic median assumptions. It does not include any of the effects from the underlying uncertainty in our assumptions or that competition might impose. This NPV is unburdened by risks.

The NPV calculation is so important to financial analysis and so repeatedly used that we should make a function out of it and keep it in a functions file that we would import every time we start a new analysis, just like we did at the beginning of this analysis with the data file.

```
CalcNPV = function(series, time, dr, eotp = TRUE) {
# series = the cash flow series
# time = index over which series is allocated
# dr = discount rate
# eotp = end of time period calculation (default is TRUE)
# or beginning of time period calculation (FALSE)

this.npv <- sum(series / (1 + dr) ^ (time - (1- eotp)))
}
```

Using this function, we would write the following to R with our currently defined assumptions.

```
npv <- CalcNPV(cash.flow, year, kDiscountRate / 100)
```

Remember that this business case deals with a project spanning two decades. The likelihood that the company will use the same discount rate over the entire horizon of concern here is small, as the company might take on or release debt, face changing interest rates on debt, experience changing volatility for the price of similar equity, and so on. More realistically, then, we might consider using an evolving discount rate for projects with long horizons. That would require stochastic simulation, which is more of a topic for Chapter 3. If a question about the importance of the discount rate arises for whether a project is worth pursuing, we can always test our sensitivity to rejecting or accepting a project on the sensitivity of the NPV to the potential evolution of the discount rate.

As a bonus, we can also find the internal rate of return (IRR) of the cash flow. The IRR is the discount rate that gives an NPV of $0. However, there are no known methods of finding the IRR in a closed form manner, so we generally use a while routine to find that value.

```
CalcIRR = function (series, time, irr0 = 0.1, tolerance = 0.00001) {
#Calculates the discount rate that produces a 0 NPV
# of a series of end-of-year cash flows.
# series = a vector of cash flows
# time = index over which series is allocated
# irr0 = the initial guess
# tolerance = the error around 0 at which the goal seek stops

my.irr <- c(irr0)
my.npv <- CalcNPV(series, time, my.irr[1])
i <- 1
while ( abs(my.npv[i]) > tolerance ) {
    if (i ==1) {
      if ( my.npv[i] > 0 ) {
        (my.irr[i + 1] = my.irr[i] * 2)
      } else {
        (my.irr[i + 1] <- my.irr[i]/2)
        }
      } else {
  # uses Newton-Raphson method of convergence
        slope.my.npv <- (my.npv[i] - my.npv[i - 1]) /
        (my.irr[i] - my.irr[i - 1])
        my.irr[i + 1] <- my.irr[i] - my.npv[i] / slope.my.npv
      }
    my.npv[i + 1] <- CalcNPV(series, time, my.irr[i + 1])
    i <- i + 1
    }
  my.irr <- my.irr[i]
}
```

Note that the `CalcIRR()` function calls the `CalcNPV()`. Therefore, we need to make sure to include the `IRR()` function in our functions file somewhere after the `CalcNPV()` function.

Be aware that the IRR most likely won't have a unique solution if the cash flow switches over the $0 line multiple times. Regardless of the several cautions that many people offer for using IRR, it is a good value to understand if the conditions for finding it are satisfactory. Namely, by comparison to the discount rate, it tells us the incremental rate by which value is created (i.e., IRR > 0) or destroyed (i.e., IRR < 0).

Using the `CalcIRR` function now in our model, we write

```
irr <- CalcIRR(cash.flow, year)
```

and get back `0.1416301`.

The IRR at 14.2% is better than our discount rate, 2.2% more, which should not surprise us because we have a positive NPV.

Write a Pro Forma Table

To create a pro forma table, we'd like to have all of the elements from the GAAP flow described earlier with the names of the GAAP elements in a column on the left, and the values for each element in a column by year.

The first thing we do is create an array composed of the pro forma elements, remembering to place a negative sign in front of the cost values.

```
pro.forma.vars <- array( c(sales, revenue, -gsa, gross.profit, -fixed.cost,
  -var.cost, -opex, -depr, op.profit.before.tax, -tax, op.profit.after.tax,
  depr, -capex, cash.flow ), dim = c(kHorizon, 14))
```

The 14 in the `dim` parameter refers to how many pro forma elements we put in the array.

We then assign that array to a data frame from the existing calculations.

```
pro.forma <- data.frame(pro.forma.vars)
```

This gives the table shown in Figure 2-11.

	X1	X2	X3	X4	X5	X6	X7	X8	X9	X10	X11	X12	X13	X14
1	0e+00	0.0e+00	0.0e+00	0	0	0	0	0	0	0.0	0	0	-2e+07	-20000000
2	0e+00	0.0e+00	0.0e+00	0	0	0	0	0	0	0.0	0	0	-2e+07	-20000000
3	2e+06	1.2e+07	-2.4e+06	9600000	-3000000	-7000000	-10000000	-5714286	-6114286	2323428.6	-3790857	5714286	-1e+07	-8076571
4	4e+06	2.4e+07	-4.8e+06	19200000	-3090000	-13300000	-16390000	-7142857	-4332857	1646485.7	-2686371	7142857	-1e+07	-5543514
5	6e+06	3.6e+07	-7.2e+06	28800000	-3182700	-18952500	-22135200	-8571429	-1906629	724518.9	-1182110	8571429	-2e+06	5389319
6	8e+06	4.8e+07	-9.6e+06	38400000	-3278181	-24006500	-27284681	-8857143	2258176	-858106.9	1400069	8857143	-2e+06	8257212
7	1e+07	6.0e+07	-1.2e+07	48000000	-3376526	-28507719	-31884245	-9142857	6972898	-2649701.1	4323197	9142857	-2e+06	11466054
8	1e+07	6.0e+07	-1.2e+07	48000000	-3477822	-27082333	-30560155	-9428571	8011274	-3044283.9	4966990	9428571	-2e+06	12395561
9	1e+07	6.0e+07	-1.2e+07	48000000	-3582157	-25728216	-29310373	-9714286	8975341	-3410629.7	5564712	9714286	-2e+06	13278997
10	1e+07	6.0e+07	-1.2e+07	48000000	-3689622	-24441805	-28131427	-4285714	15582859	-5921486.3	9661372	4285714	-2e+06	11947087
11	1e+07	6.0e+07	-1.2e+07	48000000	-3800310	-23219715	-27020025	-3142857	17837118	-6778104.7	11059013	3142857	-2e+06	12201870
12	1e+07	6.0e+07	-1.2e+07	48000000	-3914320	-22058729	-25973049	-2000000	20026951	-7610241.4	12416710	2000000	-2e+06	12416710
13	1e+07	6.0e+07	-1.2e+07	48000000	-4031749	-20955793	-24987542	-2000000	21012458	-7984734.0	13027724	2000000	-2e+06	13027724
14	1e+07	6.0e+07	-1.2e+07	48000000	-4152702	-19908003	-24060705	-2000000	21939295	-8336932.2	13602363	2000000	-2e+06	13602363
15	1e+07	6.0e+07	-1.2e+07	48000000	-4277283	-18912603	-23189886	-2000000	22810114	-8667843.4	14142271	2000000	-2e+06	14142271
16	1e+07	6.0e+07	-1.2e+07	48000000	-4405601	-17966973	-22372574	-2000000	23627046	-8978421.9	14649004	2000000	-2e+06	14649004
17	1e+07	6.0e+07	-1.2e+07	48000000	-4537769	-17068624	-21606393	-2000000	24393607	-9269570.5	15124036	2000000	-2e+06	15124036
18	1e+07	6.0e+07	-1.2e+07	48000000	-4673902	-16215193	-20889095	-2000000	25110905	-9542143.8	15568761	2000000	-2e+06	15568761
19	1e+07	6.0e+07	-1.2e+07	48000000	-4814119	-15404433	-20218553	-2000000	25781447	-9796950.0	15984497	2000000	-2e+06	15984497
20	1e+07	6.0e+07	-1.2e+07	48000000	-4958543	-14634212	-19592755	-2000000	26407245	-10034753.2	16372492	2000000	-2e+06	16372492

Figure 2-11. *The raw pro forma table*

Notice that R has oriented our pro forma with the result values in column form, and it uses a sequentially numbered X.n for the pro forma elements. Obviously, this is a little unsightly and uninformative, so we might want to replace the default column headers with header names we like better. Furthermore, we might prefer to transpose the table so that pro forma elements are on the left as opposed to across the top.

To accomplish this we first define a vector with names we like for columns.

```
pro.forma.headers <- c("Sales [lbs]", "Revenue", "GS&A", "Gross Profit",
"Fixed Cost", "Variable Cost", "OPEX", "-Depreciation", "Operating Profit
Before Tax", "Tax", "Operating Profit After Tax", "+Depreciation", "CAPEX",
"Cash Flow")
```

Next, we coerce the default headers to be reassigned to the preferred names.

```
colnames(pro.forma) <- pro.forma.headers
rownames(pro.forma) <- year
```

Then, using the transpose function t(), we reassign the pro forma to a transposed form with the column headers now appearing as row headers.

```
pro.forma = t(pro.forma)
```

A few columns of the finished form look like the table in Figure 2-12.

	1	2	3	4	5	6	7	8
Sales [lbs]	0e+00	0e+00	2000000	4000000	6000000.0	8000000.0	10000000	10000000
Revenue	0e+00	0e+00	12000000	24000000	36000000.0	48000000.0	60000000	60000000
GS&A	0e+00	0e+00	-2400000	-4800000	-7200000.0	-9600000.0	-12000000	-12000000
Gross Profit	0e+00	0e+00	9600000	19200000	28800000.0	38400000.0	48000000	48000000
Fixed Cost	0e+00	0e+00	-3000000	-3090000	-3182700.0	-3278181.0	-3376526	-3477822
Variable Cost	0e+00	0e+00	-7000000	-13300000	-18952500.0	-24006500.0	-28507719	-27082333
OPEX	0e+00	0e+00	-10000000	-16390000	-22135200.0	-27284681.0	-31884245	-30560155
-Depreciation	0e+00	0e+00	-5714286	-7142857	-8571428.6	-8857142.9	-9142857	-9428571
Operating Profit Before Tax	0e+00	0e+00	-6114286	-4332857	-1906628.6	2258176.1	6972898	8011274
Tax	0e+00	0e+00	2323429	1646486	724518.9	-858106.9	-2649701	-3044284
Operating Profit After Tax	0e+00	0e+00	-3790857	-2686371	-1182109.7	1400069.2	4323197	4966990
+Depreciation	0e+00	0e+00	5714286	7142857	8571428.6	8857142.9	9142857	9428571
CAPEX	-2e+07	-2e+07	-10000000	-10000000	-2000000.0	-2000000.0	-2000000	-2000000
Cash Flow	-2e+07	-2e+07	-8076571	-5543514	5389318.9	8257212.1	11466054	12395561

Figure 2-12. *The pro forma table formatted now with row names and time column headers*

Conduct Deterministic Sensitivity Analysis

Deterministic sensitivity represents the degree of leverage a variable exerts on one of our objective functions, like the NPV. By moving each of our variables across a range of variation, we can observe the relative strength each variable exerts on the objective. However, there is an extreme danger of relying on this simplistic type of sensitivity analysis, as it can lead one to think the deterministic sensitivity is even close to the likely sensitivity. It most likely will not be if it uses ±x% type variation that is too common in business case analysis.

Understand that the primary reason for conducting the deterministic analysis at this point is mainly to confirm the validity of the logic of the model's code, not necessarily to draw deep conclusions from it just yet. Our primary goal is to ensure that the the objective function operates directionally as anticipated.[3] For example, if price goes up, and we don't use conditional logic that makes revenue do anything other than go up, too, then we should see the NPV go up. If the construction duration lasts longer, we should see the NPV go down due to the time value of money. We really should not draw deep conclusions from our analysis until we consider all the deeper effects of uncertainty and risk, which we'll discuss in Chapter 3.

---

[3]We should on occasion be surprised by the results of a model, creating an opportunity for us to explore unanticipated sources of value and risk. A model that tells us exactly what we anticipate is redundant. A model that produces results that are beyond credibility is probably not a good model. A good model needs to be logically valid, yet produce results that give us interesting, unanticipated insights.

The best way to conduct deterministic sensitivity analysis would be to use the low and high values associated with the range obtained from SME elicitations for each variable that will eventually be treated as an uncertainty. Although the uncertain sensitivity analysis is a little more involved than the deterministic type, it does build on the basis of the deterministic analysis. In Chapter 3, not only will we use the full range of potential variation supplied by SMEs for all the variables (including those that explore the effects of competition), but each variable will run through its full range of variation as we test each sensitivity point.

Before we set up the sensitivity analysis, recall that after we read the data from the CSV file, we stored the variables' p50 values in a vector named d.vals. We'd like to do this:

1. Iterate across the d.vals vector.

2. Replace each value with a variant based on the range of sensitivity.

3. Calculate the variant NPV.

4. Store the variant NPV in an array that has as many rows as variables and as many columns as sensitivity points.

5. Replace the tested variable with its original value.

In pseudocode, this might look like this:

```
# iterate across our list of variable values
for (i in index of d.vals) {
# iterate across the list of sensitivity points
    for (k in index of sensitivity points) {

# replace the variable fixed value with each sensitivity point
        d.vals[i] <- d.vals.sens.points[i, k]

# calculate the NPV
        npv <- {...}

# record the value of the movement of the NPV
        npv.sens[i, k] <- npv
    }

# restore the current variables' base value
    d.vals[i] <- d.vals.original[i]
}
```

The way we implement this in real R code is to duplicate the original R file. Then we wrap the for loops around the original code. Before we do this, though, we set up parameters that control the looping after the variable values are assigned to the d.vals vector and before the individual assignment of value to each variable.

```
# Slice the values from data frame d.data.
d.vals <- d.data$p50[1:13]

d.vals.sens.points <- d.data[1:13, 2:4]
sens.point <- 1:3
len.d.vals <- length(d.vals)
len.sens.range <- length(sens.point)

# Sets up an initialized 14x3 array of 0s.
  npv.sens <- array(0, c(len.d.vals, len.sens.range))
```

The for loop code blocks surround our original code that found the NPV.

```
 for (i in 1:len.d.vals) {
   for (k in 1:len.sens.range) {
     d.vals[i] <- d.vals.sens.points[i, k]

        # Assign values to variables.
        ...
        npv.sens[i, k] <- npv
     }
     d.vals[i] <- d.data$value[i]
   }
}
```

The resultant npv.sens matrix looks like this.

```
          [,1]     [,2]        [,3]
[1,]   11451979 8214363    3681699.6
[2,]   13077695 8214363     288329.9
[3,]   10748136 8214363    3146815.1
[4,]    6684537 8214363   10858019.4
[5,]    9938517 8214363    4766053.3
[6,]   11421355 8214363    1800378.4
[7,]    8786129 8214363    6949780.0
```

```
[8,]    25703301 8214363 -26763514.8
[9,]     2357020 8214363  18256935.0
[10,]   9888497 8214363   4866094.4
[11,]  11656823 8214363   1193338.6
[12,]  -7779381 8214363  40201850.8
[13,]-2526832  08214363 75179728.2
```

To make this table more understandable, we can convert it to a data.frame with
column names equal to the sensitivity points and the row names equal to the variable
identifiers that are driving the NPV using the following coercions.

```
var.names <- d.data[1:13, 1]
sens.point.names <- c("p10", "p50", "p90")
rownames(npv.sens) <- var.names
colnames(npv.sens) <- sens.point.names
```

The result is a table for npv.sens that looks like this.

	p10	p50	p90
p1.capex	11451979	8214363	3681699.6
p1.dur	13077695	8214363	288329.9
p2.capex	10748136	8214363	3146815.1
p2.dur	6684537	8214363	10858019.4
maint.capex	9938517	8214363	4766053.3
fixed.prod.cost	11421355	8214363	1800378.4
prod.cost.escal	8786129	8214363	6949780.0
var.prod.cost	25703301	8214363	-26763514.8
var.cost.redux	2357020	8214363	18256935.0
gsa.rate	9888497	8214363	4866094.4
time.to.peak.sales	11656823	8214363	1193338.6
mkt.demand	-7779381	8214363	40201850.8
price	-25268320	8214363	75179728.2

The way we interpret this table is that each value is the NPV for our business
case when each variable is independently varied through their p10, p50, p90 values
independently of the others. For example, if the price were to go to its low (or p10)
value, the NPV would shift to -$25.3 million. On the other hand, if the price were to go
to its high value, the NPV would shift to $75.2 million. Accordingly, the price, market

43

demand, and variable production costs are the factors that could cause the NPV to go
negative or financially undesirable. If we delude ourselves into thinking there will be no
competition, the other variables don't seem to cause enough change to worry about.
We'll leave the analysis to the risk layer of the model that accounts for uncertainties
associated with the competitive effects.

The following code produces the sensitivity chart using the barplot() function.

```
npv.sens.array <- array(0, c(len.d.vals, 2))
npv.sens.array[, 1] <- (npv.sens[, 1] - npv.sens[, 2])
npv.sens.array[, 2] <- (npv.sens[, 3] - npv.sens[, 2])
rownames(npv.sens.array) <- var.names
colnames(npv.sens.array) <- sens.point.names[-2]

# Calculates the rank order of the NPV sensitivity based on the
# absolute range caused by a given variable. The npv.sens.array
# is reindexed by this rank ordering for the bar plot.
npv.sens.rank <- order(abs(npv.sens.array[, 1] - npv.sens.array[, 2]),
decreasing = FALSE)

ranked.npv.sens.array <- npv.sens.array[npv.sens.rank, ]
ranked.var.names <- var.names[npv.sens.rank]
rownames(ranked.npv.sens.array) <- ranked.var.names

par(mai = c(1, 1.75, 0.5, 0.5))
barplot(t(ranked.npv.sens.array) / 1000,
            main = "Deterministic NPV Sensitivity",
            names.arg = ranked.var.names,
            col = "light blue",
            xlab = "NPV [$000]",
            beside = TRUE,
            horiz = TRUE,
            offset = npv.sens[, 2] / 1000,
            las = 1,
            space = c(-1, 1),
            cex.names = 1,
            tck = 1)
```

The result of the code graphically reveals the information in our sensitivity table. The inherent pattern in the numbers, which remains difficult to discern in its tabular form, becomes immediately obvious in Figure 2-13. We typically refer to this kind of chart as a *tornado chart.*

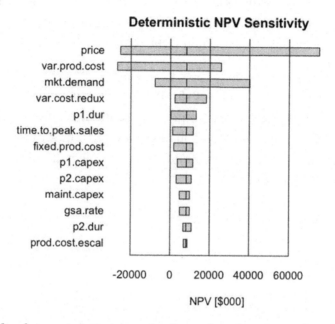

Figure 2-13. *The deterministic tornado chart showing the ordered sensitivity of the NPV to the range of inputs used for the uncertainties*

Although we can look up the meaning of the parameters for the `barplot()` function in R Help, I think it's worth commenting on the last few parameters in the function as used here.

- The `beside = TRUE` parameter sets the bar plot columns adjacent to each other as opposed to being stacked (i.e., `beside = FALSE`).

- `horiz = TRUE` sets the tornado chart in a vertical orientation because the data frame for `npv.sens.array` has the names of the variables in the position that `barplot()` treats as the x axis.

- `offset` moves the baseline of the bar plot, which is 0 by default, to some desired bias position. In our tornado chart, we want the baseline to be the NPV of the middle value of the sensitivity points.

- `las = 1` sets the names of the variables to a horizontal orientation.

- `space = c(-1,1)` makes the bars in a row overlap. Namely, the -1 moves the upper bar in a row down by the distance of its width.

- `cex.names = 1` sets the height of the names of the variables to a percentage of their base height. I usually leave this in place so that I can quickly modify the value to test the visual aesthetics of changing their size.

Finally, the `par(mai = c(1, 1.75, 0.5, 0.5))` statement, which we placed before the `barplot` function, sets the width of the graphics rendering window to accommodate the length of the names of the variables. You might want to adjust the second parameter of this term, here set to 1.75, to account for different variable names if you choose to expand the complexity of this model.

## CHAPTER 3

Include Uncertainty in the Financial Analysis

We incorporate potential effects of uncertainty in our analysis to account for the fact that we don't know what will happen in the future. Monte Carlo simulation is the method that allows us to include these considerations of uncertainty analytically.

Why and How Do We Represent Uncertainty?

Looking forward to future outcomes that are important to consider in planning exercises, we realize rather quickly that we rarely know the exact representative values of those outcomes. We don't know exactly how much something will cost, or how long it will take to accomplish a set of tasks, or what the actual adoption rate will be for a given product. The list goes on. Just because we don't know what the actual outcomes will be, however, doesn't give us license to avoid considering them in our planning. On the other hand, when we consider all of the multiple ways in which the various uncertain outcomes can be combined, the number of calculations required becomes intractably large. Most people recognize this, and they try to overcome the limitations with the tools they use for analysis (i.e., spreadsheets), or through their knowledge of how to treat the uncertainty via workarounds, usually in at least two steps.

The first step many people take to handle the problem of uncertainty is to try estimating most likely cases for the appropriate assumptions in their planning, or to consider worst or best case scenarios in an effort to consider conservative positions. What they usually fail to consider are the numerous ways in which bias creeps into their estimates of most likely outcomes (indeed, they might not even be aware of the subtle ways it happens), or the number of improbable conditions that are required to justify the particular caveats that define best or worst case scenarios.

R. D. Brown III, *Business Case Analysis with R*, https://doi.org/10.1007/978-1-4842-3495-2_3

The second step people take attempts to account for the effect of ranges around the assumptions on the objective function. Here again, bias creeps in because the range of sensitivity chosen is not usually tied to a likely range of variation. Instead, people use arbitrary ranges of variation, like ±20%, across all their assumptions as a way to accommodate what they think is a significant range.

Of course, there are also those analysts with a masochistic bent who attempt a brute force approach of testing dozens of scenarios sequentially, convincing themselves they are gaining a comprehensive and detailed understanding of their problem. All they are really doing is getting mired in analysis paralysis. The temptation to test "just one more scenario" never ends. The worst case of this mindset I have ever observed was a client who had produced 216 scenarios of a capital expansion plan over the course of two years, and they were no closer to making meaningful decisions than the day they began. Such wasted time only leads to value attrition and missed opportunity.

The net result of the first two steps is that people accommodate uncertainty with biased rationalizations, falsely communicating an unwarranted degree of precision by the single point numbers chosen in their analysis. Then they "stress test" their analysis with arbitrarily chosen relative ranges on assumptions. Bias, false precision, and arbitrariness don't seem like the basis for clear and valid analysis to me.

The answer to this problem is actually similar to the brute force approach; that is, we consider a large number of scenarios across the range of likely outcomes of the assumptions. However, instead of performing our calculations sequentially and only on the scenario combinations we tend to think matter most, we perform the calculations in parallel and according to the coherent and consistent rules of probability. This means that we must get used to thinking in ranges of outcomes (not single points) and corresponding degrees of likelihood across each assumption's range. In short, we should treat uncertain assumptions as probability distributions. That's the notional side of the effort. The functional side incorporates a process for conducting the calculations called *Monte Carlo simulation*.

How we actually gather information to represent assumptions as probability distributions and what that even means is presented in Part 3 of this book, which presents a process for eliciting distributions from SMEs. The following resources also provide excellent guidance in this area.

- Peter MacNamee and John Celona, *Decision Analysis for the Professional* (San Jose, CA: SmartOrg, 2007).

- Gregory S. Parnell, *Handbook of Decision Analysis* (Hoboken, NJ: Wiley, 2013).

- Patrick Leach, *Why Can't You Just Give Me the Number?: An Executive's Guide to Using Probabilistic Thinking to Manage Risk and Make Better Decisions* (Sugar Land, TX: Probabilistic Publishing, 2014).

What Is Monte Carlo Simulation?

Before I delve into the technical details of Monte Carlo simulation, I provide a little analogy to help clarify the essential mechanism by which it operates. Imagine a BB gun (the Red Ryder Carbine Action 200-shot Range Model air rifle, to be exact) mounted to a slider on a vertical rail with a fixed height. The BB gun always points horizontally at a 90° angle to the vertical rail. When it fires a BB, the trajectory of the BB is perfectly flat until it hits a special obstacle that has been erected at a fixed distance from the vertical rail. The obstacle is like a wall, but it leans over at an angle such that its vertical height is exactly the same as the vertical rail that holds the BB gun, and it's shaped a bit like a French curve.

The wall also has a special property so that when a BB hits it, the BB passes through it but falls down vertically to the ground. Underneath this wall are little cups sitting along the length of the wall such that each cup touches its adjacent cup, and each cup is numbered in sequence starting at some lowest value up to a highest value with equal step sizes. Now that we've accepted this setup, fantastic as it sounds, imagine that a gremlin randomly moves the BB gun up and down the vertical rail, firing a BB at each position that it selects. Over time, we notice that the cups underneath the wall are filling up with BBs, and the most BBs fill the cups closest to the section of the wall that is most vertical (see Figure 3-1).

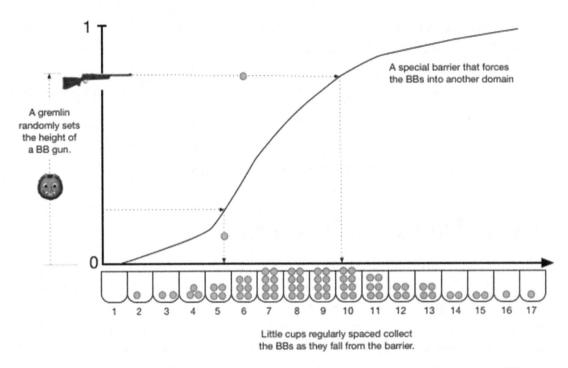

Figure 3-1. *Monte Carlo simulation is like a gremlin shooting BBs at a wall from a random vertical postion. The wall records the horizontal position of where it was struck by dropping the BB into a cup underneath.*

The process describing how our gremlin randomly shoots BBs at a wall is almost exactly how Monte Carlo simulation of a random variable works, except that instead of BBs being fired from a random vertical position, a computer selects random numbers strictly in the interval from 0 to 1. Each of these random numbers gets mapped to another value with a function that can produce numbers in any other arbitrary domain. For example, suppose we want to re-create a possible sample distribution of adult humans in a situation in which their heights are of interest to us. If we had a function that describes the distribution of the heights of the population of interest, we could randomly create a test population of their heights without any prejudice toward one side of the range or another.

Let's take our analogy a little further. Imagine now that we play a game with of these shooting ranges, each with its own gremlin and uniquely shaped function wall. Let's denote the function walls with the names A, B, and C. Our gremlins are coordinated such that they each fire their BB guns at the same time, but they independently choose the vertical position from which they fire. As it turns out, we are given a numerical relationship between A, B, and C, which is Y = A * (B + C). On each round of firing, we note the value of the cups the BBs fall into from their respective ranges, we perform the calculation A * (B + C), and we record the value of Y. The game we are playing is to try to predict where most of the values of all possible values of Y will occur after we've observed the gremlins shoot a sufficiently large—but not exhaustively large—number of BBs.

This situation is not as silly as it sounds. Suppose we want to estimate the range and prevalence of most of our costs for manufacturing berets, raspberry and lemon colored, and the manufacturing cost could vary over a range during the time we receive orders for berets, the actual number of which we won't know until it's time to close the books. Our formula for the exact cost of manufacturing in our period of concern will be mfg.cost = unit.mfg.cost * (number.raspberry + number.lemon).

The potential range and prevalence of the final manufacturing cost could be simulated using the gremlin method or we could use Monte Carlo simulation to simulate the range of possible values of the terms of our cost formula as long as we keep track of the outcomes of each of the terms. Once we've done such a thing, we would be in a better position to more accurately budget our operational costs for the year or write favorable contingent contracts with our suppliers if unit costs exceeded a certain value. The Monte Carlo simulation process gives us the ability to gain better clarity about how the world might turn out so that we can plan and design more desirable situations for ourselves and others.

A probability distribution for an uncertain assumption is defined within some real number domain and over a specific interval within that domain. Within the specific interval, probabilities are assigned to nonoverlapping subintervals that express our degrees of belief that the subintervals can occur or long-run frequencies that the subintervals will occur. We must conform to the inviolate principle that the sum of the probabilities equals exactly 1 within the interval.

For example, a cost for a project task might be defined in the domain from $0 to infinity, but the potential interval denoted by the context of the problem at hand might only cover $100,000 to $500,000. Within that interval, probabilities are assigned either in discrete buckets (e.g., Prob($100,000 < X <= $200,000) = 0.1; Prob($200,000 < X <= $300,000) = 0.25; Prob($300,000 < X <= $400,000) = 0.5; Prob($400,000 < X <= $500,000) = 0.15) or according to some mathematical function that assigns probabilities in a continuous form.[1]

---

[1] The interval might actually be as large as the domain itself, as is the case with a normal distribution, which spans all real numbers even though the bulk of the distribution can be assigned to exist within a relatively narrow range of dispersion.

Once these parameters are defined, the Monte Carlo algorithm randomly selects values from within an assumption's interval according to the frequency implied by its internal probabilities over hundreds to thousands of iterations. This process is repeated in parallel for all the assumptions, some of them independently and others in a coordinated, or conditional, manner as required by the business logic. On each iteration, the intermediate calculations in the business case model are then performed as if only single point values were considered. Each iteration of the simulation represents a parallel universe to the other iterations. The outcomes on each iteration are stored, and the net results are examined in bulk. Consequently, through the combination of the business logic and the probabilities defined on the assumptions' intervals, we can get a much less biased view of what the most likely range of conditions will be for our objective function. Of course, our assumptions' definitions might be biased from the beginning. The books mentioned earlier describe methods for controlling for that condition, as do later sections of this book.

The Matrix Structure of Business Case Simulation in R

With the exception of the depreciation array, we based the deterministic model framework in one-dimensional vectors, as all the calculations produced results along a single index, year. To add the risk layer of the model, we need to add another dimension that handles the parallel scenarios. This dimension, run, will exist orthogonally to the year index. Its length, the number of iterations we choose to run (kSampsize), will be the same consistently for every calculation.

The following is an example of the first few rows of the phase calculation using uncertain values for the Phase 1 duration (p1.dur) and Phase 2 duration (p2.dur).

```
> phase
       [,1] [,2] [,3] [,4] [,5] [,6] [,7] [,8] [,9] [,10] [,11] [,12] [,13] [,14] [,15] [,16] [,17] [,18] [,19] [,20]
 [1,]    1    1    1    1    1    2    2    2    2     3     3     3     3     3     3     3     3     3     3     3
 [2,]    1    1    1    2    2    3    3    3    3     3     3     3     3     3     3     3     3     3     3     3
 [3,]    1    1    1    1    2    2    2    3    3     3     3     3     3     3     3     3     3     3     3     3
 [4,]    1    1    1    1    2    2    2    3    3     3     3     3     3     3     3     3     3     3     3     3
 [5,]    1    1    2    2    3    3    3    3    3     3     3     3     3     3     3     3     3     3     3     3
 [6,]    1    1    1    1    1    2    2    2    3     3     3     3     3     3     3     3     3     3     3     3
 [7,]    1    1    1    2    2    2    3    3    3     3     3     3     3     3     3     3     3     3     3     3
 [8,]    1    1    2    2    2    3    3    3    3     3     3     3     3     3     3     3     3     3     3     3
 [9,]    1    1    1    1    2    2    2    3    3     3     3     3     3     3     3     3     3     3     3     3
[10,]    1    1    1    1    1    2    3    3    3     3     3     3     3     3     3     3     3     3     3     3
[11,]    1    1    1    1    1    2    2    2    2     3     3     3     3     3     3     3     3     3     3     3
[12,]    1    1    1    1    2    3    3    3    3     3     3     3     3     3     3     3     3     3     3     3
[13,]    1    1    1    2    3    3    3    3    3     3     3     3     3     3     3     3     3     3     3     3
[14,]    1    1    1    1    1    1    2    3    3     3     3     3     3     3     3     3     3     3     3     3
[15,]    1    1    1    2    2    2    3    3    3     3     3     3     3     3     3     3     3     3     3     3
```

Notice that the `year` index elements produce the columns of the matrix, and the `run` index elements produce the rows. Each row represents a potential manifestation of the world in which a specific set of outcomes are realized for the Phase 1 and Phase 2 durations.

Before I explain how to set up this kind of matrix, I discuss some ways in which we can generate subjective distributions for each of our assumptions that represent the uncertainty we might face.

Useful Distributions for Expert Elicited Assumptions

Business case analysis often depends on making assumptions for which we possess little or no documented data from which we can develop appropriate empirical distributions. As I mentioned earlier, though, not having so-called hard data doesn't excuse us from using a distribution. In these cases, we need to extract the information we need from SMEs and encode their knowledge, expertise, and degrees of uncertainty into a procedural function. I describe here four useful approaches for using expert elicited values from SMEs to produce distributions for uncertainties.

Discrete Distributions: McNamee-Celona and Swanson-Megill

In the early days of the decision analysis program, tractable ways to simulate uncertainty were handled by the Gaussian quadrature, McNamee-Celona,[2] and later, the extended Swanson-Megill[3] methods. These methods provide discrete approximations of single-mode continuous distributions in some domain of concern. Specifically, they derive three probability weightings that specify the frequency of discrete outcomes representing an expert's 80th percentile prediction interval values that preserves the mean and variance of the continuous distributions implied by the expert elicitation. Gaussian quadrature or McNamee-Celona were used for symmetric distributions; extended Swanson-McGill was used for skewed distributions. The three weights are associated with the 10th, 50th, and 90th cumulative probability quantiles.

---

[2]Peter McNamee and John Celona, *Decision Analysis with Supertree* (2nd ed., San Francisco: The Scientific Press, 1990).

[3]R. E. Megill, *An Introduction to Risk Analysis* (2nd ed., Tulsa, OK: PennWell, 1984).

This simplification of continuous variable outcomes permitted a cheaper means of running a simulation when numerical processing was much more expensive than it is today. In fact, the simplification could be conveniently performed by calculator and hand-drawn decision trees.

The three methods typically use the following weights, or probabilities, for each of three discrete intervals in the represented event.

- Gaussian quadrature, McNamee-Celona: Prob(X10, X50, X90) = (0.25, 0.5, 0.25)

- Extended Swanson-Megill: Prob(X10, X50, X90) = (0.3, 0.4, 0.3)[4]

For example, let's say we interview an SME for the potential range of outcomes for the annual production costs. Our elicitation yields the following values:

- ProductionCosts10 = $2.75 million

- ProductionCosts50 = $3.00 million

- ProductionCosts90 = $3.75 million

These values reveal that, given the best information available to our SME, he believes that there is a 10% chance the outcome will be less than $2.75 million, and a 90% chance the outcome will be less than $3.75 million. He places even odds that the outcome could be higher or lower than $3.00 million. When we develop a simulation of values in this range, our algorithm should select the ProductionCosts10 = $2.75 million approximately 30% of the time; ProductionCosts50 = $3.00 million 40% of the time; and ProductionCosts90 = $3.75 million 30% of the time. The extended Swanson-Megill weightings are chosen because the range is skewed around ProductionCosts50.

Be aware that using a discrete method poses some problems, namely that it sacrifices accuracy for speed and ease of use. When the discrete method is used for many independent variables in a simulation, the loss of accuracy is minimized somewhat; however, if a simulation relies on only a few distributions represented by these discrete approximations, the loss of accuracy can be pronounced.

---

[4]There are several other methods that have been proposed for simulating distributions with discrete weightings for three distinct values. All of them essentially use the same approach described here.

The following code provides the McNamee-Celona and Swanson-Megill approach.

```
CalcMCSM = function(p10, p50, p90, samples) {
# This function simulates a distribution from three quantile
# estimates for the probability intervals of the predicted outcome.
# p10 = the 10th percentile estimate
# p50 = the 50th percentile estimate
# p90 = the 90th percentile estimate
# samples = the number of runs used in the simulation
# Note, this function strictly requires that p10 < p50 < p90.
# If the distribution is skewed, the quantiles are selected with
# frequency:
# p10 -> 0.3, p50 -> 0.4, and p90 -> 0.3 from Swanson-Megill method.
# If the distribution is symmetric, the quantiles are selected with
# frequency p10 -> 0.25, p50 -> 0.5, and p90 -> 0.25 from
# McNamee-Celona method. The process of simulation is simple Monte
# Carlo. The returned result is a vector containing the values of the
# quantiles at frequencies represented by the method indicated by the skew.

  if (p10 >= p50 || p50 >= p90) {
    stop("Parameters not given in the correct order. Must be
    given as p10 < p50 < p90.")
  } else {
# Determine if the distribution is skewed or not.
    skew <- (p50 - p10 != p90 - p50)

# Create a uniform variate sample space in the interval (0,1).
    U <- runif(samples, 0, 1)

# Select the quantiles at the frequency indicated by the skew.
    X <- if (skew == T) {
      (U <= 0.3) * p10 + (U > 0.3 & U <= 0.7) * p50 + (U > 0.7) * p90
    } else {
      (U <= 0.25) * p10 + (U > 0.25 & U <= 0.75) * p50 + (U > 0.75) * p90
    }
    return(X)
  }
}
```

Using the values in the production costs example earlier gives the cumulative probability distribution in Figure 3-2 after some postprocessing on the returned sample values.

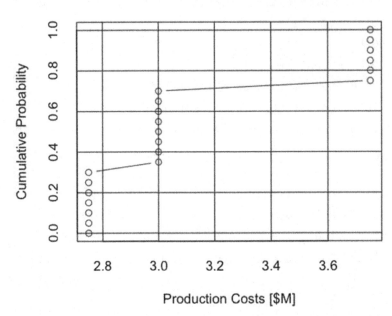

Figure 3-2. *The cumulative probability distribution of the discrete approximation of the continuously distributed production costs using the extended Swanson-Megill generator*

Continuous Distributions

Triangular

The triangular distribution is a distribution frequently used for the simulation of uncertainties that preserves the continuous nature of the represented event. In this regard, the triangular is an incremental improvement over the discrete methods. Like the discrete methods, the distribution requires three parameters: a low estimate, a mode estimate, and a high estimate.

I rarely use this distribution (or its smoother close cousin, the Beta-PERT) unless the SME I consult knows specifically what the two extreme values are, usually defined by some constraint in the underlying system, and the mode (the most likely outcome across the distribution) is known with strong empirical evidence. I follow this guideline because my own experience has shown me that SMEs are not very good at estimating most likely outcomes (regardless of the distribution being considered) without importing a number of extremely powerful biases; we explore ways to deal with this in later sections of this book.

Biases arise from many cognitive and motivational drivers, including these:

- *Availability:* Recalling values that are memorable or easily accessible.

- *Anchoring:* Using the first best guess as a starting point for subsequent estimating.

- *Expert overconfidence:* Failure of creativity or hubris (e.g., "I know this information and can't be wrong because I'm the expert").

- *Incentives:* The SME experiences some benefit or cost in relationship to the outcome of the term being measured, adjusting the estimate in the direction of the preferred outcome.

- *Entitlement:* The SME provides an estimate that reinforces his or her sense of personal value.

Therefore, if you must use a triangular distribution (or the Beta-PERT), SMEs should be aware that when they think about the low and high values, they believe the chance that any other outcome beyond these limits is nearly zero. Furthermore, they clearly understand that the most likely case is the mode, and not an average of the range of the interval or the median case (the point at which they believe there are equal odds that the outcome could either be higher or lower). The Beta-PERT distribution also suffers from some other limitations; namely, it often underestimates the uncertainty in the natural distribution (i.e., task durations) it is frequently employed to represent. For this reason, I never use the Beta-PERT, as I think it performs even less well than the triangular distribution to communicate the full effect of uncertainty on the final value function.

```
CalcTriang = function(n, l, mode, h, samponly=TRUE) {
# This function simulates a triangular distribution from three
# estimates for the interval of the predicted outcome.
# l = low value
# mode = the most likely value
# h = the high value
# n = the number of runs used in the simulation
# Note, this function strictly requires that p10 < p50 < p90.
# The process of simulation is simple Monte Carlo.
# The returned result is, by default, a vector of values for X if
```

```
# samponly=TRUE, else a (n x 2) matrix is returned such that
# the first column contains the domain samples in X, and the
# second column contains the uniform variate samples from U.

        if (l == mode && mode == h) {
                return(rep(mode, n))
        } else if (l > mode || mode > h) {
                stop ("Parameters not given in the correct order. Must be
                given
                    as l <= mode <= h.")
        } else {

#Create a uniform variate sample space in the interval (0,1).
                U <- runif(n, 0, 1)

# The relative position of the mode in the interval (l, h).
                m <- (mode - l)/(h - l)

# The lower tail quadratic solution of x in a unit space.
                x1 <- (U <= m) * sqrt(U * m)

# The upper tail quadratic solution of x in a unit space.
                x2 <- (U > m)*((-sqrt((m - U)/(1 - m) + 1) + 1)*(1 - m) + m)

# Fits the solution to the full range indicated by the inputs.
                X <- l + (h - l) * (x1 + x2)

# Return the results.
                if (samponly == TRUE) {
                return(X)
        } else {
                X <- array( c(X, U), dim=c(n, 2))
                return(X)
        }
    }
}
```

Using the values in the production costs example again gives the cumulative probability distribution in Figure 3-3, also after some postprocessing on the returned sample values.

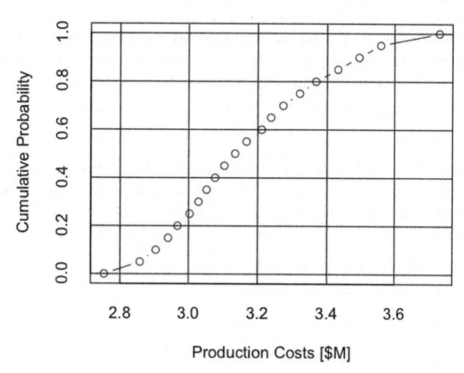

Figure 3-3. *The cumulative probability distribution of the production costs using the triangular distribution generator*

However, you should not think of p10, p50, and p90 estimates as corresponding to low, mode, and high parameters for a triangular distribution. I did that here merely to demonstrate the shape of the resultant curve.

BrownJohnson Distribution

Some years ago I became frustrated with the limitations of the discrete and triangular parameterizations, the tedious process of assessing a full distribution from an expert when dozens of assumptions had to be assessed, or finding an approximate known distribution to fit a few data points gathered from an SME. To this latter point, it just wasn't intellectually satisfying to use a distribution just because I could find a similar geometry to the data I had, regardless of whether the distribution actually matched the underlying process being assessed. Furthermore, in some sense, I am indifferent to the use of any known distributions for business case simulations (unless I have good empirical evidence and a clear rationale that convinces me otherwise), as all I really care about is representing the uncertainty as reported by an SME in a logically consistent and probabilistically coherent manner. My belief is that if the SME is properly calibrated, his or her report will conform to any underlying process as understood by his or her mind.

I derived a parametric guidance that approximates a full assessment using only the 10th, 50th, and 90th cumulative probability quantiles. Then I used a piecewise quadratic spline to build the associated cumulative probability distribution. The guidance goes as follows:

- p0 = 2.5 * p10 - 1.5 * p50

- p100 = 2.5 * p90 - 1.5 * p50

The p0 and p100 are not intended to represent the actual values beyond which no other outcome can occur. Rather, they are virtual endpoints that capture approximately 99.9% of the range of uncertainty implied by the p10, p50, and p90 with Gaussian-like tails.

If we use the values from our example in the Swanson-Megill section earlier,

- ProductionCosts10 = $2.75 million

- ProductionCosts50 = $3.00 million

- ProductionCosts90 = $3.75 million

from the formula for the p0 and p100 tail points, we get these values:

- p0 = 2.5 * $2.75 million - 1.5 * $3.00 million = $2.38 million

- p100 = 2.5 * $3.75 million - 1.5 * $3.00 million = $4.88 million

If the p0 or p100 values fall outside allowable intervals (e.g., below 0), we can truncate the distribution at the desired boundaries using something like max(low. constraint, p0) or min(high.constraint, p100).

My colleague, Eric Johnson, later replaced the piecewise quadratic spline with a piecewise linear because it was just easier to handle in this form. He also describes this approach in the book he recently coauthored, *Handbook of Decision Analysis.*[5]

The following code provides the linear spline approach.

```
CalcBrownJohnson = function(minlim=-Inf, p10, p50, p90, maxlim=Inf, n,
samponly=TRUE) {
# This function simulates a distribution from three quantile
# estimates for the probability intervals of the predicted outcome.
```

---

[5]Gregory S. Parnell PhD, Terry Bresnick MBA, Steven N. Tani PhD, Eric R. Johnson PhD, *Handbook of Decision Analysis* (Hoboken, NJ: Wiley, 2013, pp. 256-257).

p10 = the 10th percentile estimate
p50 = the 50th percentile estimate
p90 = the 90th percentile estimate
n = the number of runs used in the simulation
Note, this function strictly requires that p10 < p50 < p90.
The process of simulation is simple Monte Carlo.
The returned result is, by default, a vector of values for X if
samponly=TRUE, else a (n x 2) matrix is returned such that the first
column contains the domain samples in X, and the second column
contains the uniform variate samples from U.

```
  if (p10 == p50 && p50 == p90) {
            return(rep(p50, n))
      } else if (p10 >= p50 || p50 >= p90) {
    stop("Parameters not given in the correct order. Must be
    given as p10 < p50 < p90.")
  } else {
```

#Create a uniform variate sample space in the interval (0,1).
```
    U <- runif(n, 0, 1)
```

Calculates the virtual tails of the distribution given the p10, p50, p90
inputs. Truncates the tails at the upper and lower limiting constraints.
```
    p0 <- max(minlim, 2.5 * p10 - 1.5 * p50)
    p100 <- min(maxlim, 2.5 * p90 - 1.5 * p50)
```

#This next section finds the linear coefficients of the system of linear
equations that describe the linear spline, using linear algebra...
[C](A) = (X)
(A) = [C]^-1 * (X)
In this case, the elements of (C) are found using the values (0, 0.1,
0.5, 0.9, 1) at the endpoints of each spline segment. The elements
of (X) correspond to the values of (p0, p10, p10, p50, p50, p90,
p90, p100). Solving for this system of linear equations gives linear
coefficients that transform values in U to intermediate values in X.
Because there are four segments in the linear spline, and each
segment contains two unknowns, a total of eight equations are
required to solve the system.

```
# The spline knot values in the X domain.
   knot.vector <- c(p0, p10, p10, p50, p50, p90, p90, p100)

# The solutions to the eight equations at the knot points required to
# describe the linear system.
   coeff.vals <- c(0, 1, 0, 0, 0, 0, 0, 0, 0.1, 1, 0, 0, 0, 0, 0, 0,
   0, 0, 0.1, 1, 0, 0, 0, 0, 0, 0, 0.5, 1, 0, 0, 0, 0, 0, 0, 0, 0, 0.5,
   1, 0, 0, 0, 0, 0, 0, 0.9, 1, 0, 0, 0, 0, 0, 0, 0, 0, 0.9, 1, 0, 0, 0,
   0, 0, 0, 1, 1)

# The coefficient matrix created from the prior vector looks like the
# following matrix:
#     [, 1]    [, 2]    [, 3]    [, 4]    [, 5]    [, 6]    [, 7]  [, 8]
# [1,] 0.0     1.0      0.0      0.0      0.0      0.0      0.0    0.0
# [2,] 1.0     1.0      0.0      0.0      0.0      0.0      0.0    0.0
# [3,] 0.0     0.0      1.0      1.0      0.0      0.0      0.0    0.0
# [4,] 0.0     0.0      0.5      1.0      0.0      0.0      0.0    0.0
# [5,] 0.0     0.0      0.0      0.0      0.5      1.0      0.0    0.0
# [6,] 0.0     0.0      0.0      0.0      0.9      1.0      0.0    0.0
# [7,] 0.0     0.0      0.0      0.0      0.0      0.0      0.9    1.0
# [8,] 0.0     0.0      0.0      0.0      0.0      0.0      1.0    1.0

      coeff.matrix <- t(matrix(coeff.vals, nrow=8, ncol=8))

# The inverse of the coefficient matrix.
   inv.coeff.matrix <- solve(coeff.matrix)

# The solution vector of the linear coefficients.
   sol.vect <- inv.coeff.matrix %*% knot.vector

#Builds the response by the piecewise linear sections
   X <- (U <= 0.1) * (sol.vect[1, 1] * U + sol.vect[2, 1]) +
      (U > 0.1 & U <= 0.5) * (sol.vect[3, 1] * U + sol.vect[4, 1]) +
      (U > 0.5 & U <= 0.9) * (sol.vect[5, 1] * U + sol.vect[6, 1]) +
      (U > 0.9 & U <= 1) * (sol.vect[7, 1] * U + sol.vect[8, 1])

   if (samponly == TRUE) {
     return(X)
   } else {
```

```
    X <- array( c(X, U), dim=c(n, 2))
    return(X)
  }
 }
}
```

In this case, the resultant cumulative probability distribution in Figure 3-4 clearly demonstrates the linear spline that maps the uniform variate onto the production cost domain.

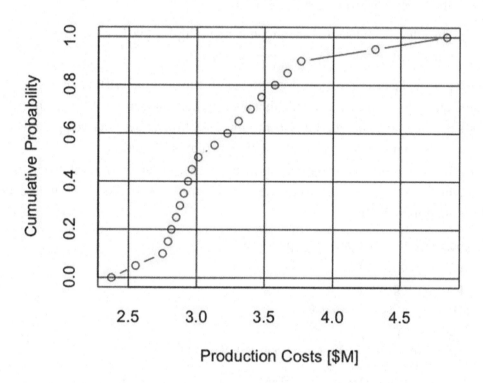

Figure 3-4. *The cumulative probability distribution of the production costs using the piecewise linear BrownJohnson generator*

Lognormal Distribution

In some cases, using a known distribution is warranted, particularly when SMEs are assessing an outcome that they understand to conform to certain kinds of underlying, systematic processes, and a large body of evidence supports the selection of a given shape. For such situations in which the following are true, you should consider using a lognormal distribution.

- Potential outcomes follow an asymmetric distribution.

- The upper tail theoretically extends indefinitely.

- The lower tail never extends below 0.

- The effects that lead to the outcome are multiplicative in nature and not merely additive (i.e., not mean-reverting).

The lognormal distribution possesses the interesting characteristic that if two estimated quantile points are available that are symmetric around a p50 (e.g., the p10 and p90), then we can simulate the distribution easily from readily derived parameters that R recognizes for use in either the rnorm() or rlnorm() functions.

The following uses p10 and p90 estimates to generate a vector of normally distributed values in the log scale before returning the values in the measured domain. It uses the rnorm() function to do this.

```
CalcLognorm80N = function(p10, p90, samples) {
     lp10 <- log(p10)
     lp90 <- log(p90)
     p50 <- p10*sqrt(p90/p10)
     lp50 <- log(p50)
     lgs <- abs(lp90 - lp50)/qnorm(0.9, 0, 1)

     X <- rnorm(samples, lp50, lgs)
     return(exp(X))
}
```

This approach simply uses the rlnorm() function.

```
CalcLognorm80L = function(p10, p90, samples) {
     lp10 <- log(p10)
     lp90 <- log(p90)
```

```
p50 <- p10*sqrt(p90/p10)
lp50 <- log(p50)
lgs <- abs(lp90 - lp50)/qnorm(0.9, 0, 1)

X <- rlnorm(samples, lp50, lgs)
return(X)
}
```

Which one you choose to use is entirely up to your discretion.

Using only the p10 and p90 values of the production costs estimates, we get the cumulative probability distributions shown in Figure 3-5 using both of the lognormal distribution functions.

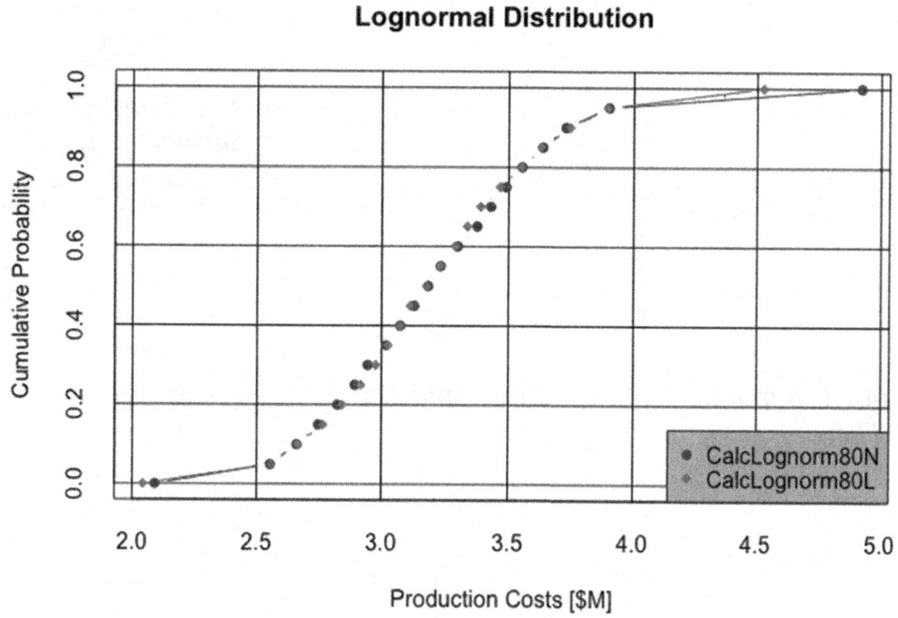

Figure 3-5. *The cumulative probability distributions of the production costs using two slightly different approaches to implement a lognormal distribution generator*

Note the high degree of coherence between them. We can attribute the small observable difference to simulation error.

When petroleum reservoir engineers use this type of function to estimate the amount of reserves in a field, they typically truncate the upper tail at the 99th percentile to accommodate the fact that a reservoir cannot be infinite in size, which the lognormal distribution would theoretically predict.

Now we compare all four distribution types in Figure 3-6.

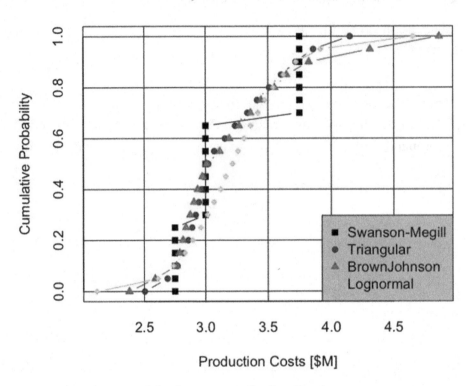

Figure 3-6. *Comparison of the four example distribution generators discussed here*

Modify the Influence Diagram to Reflect the Risk Layer

Now we need to modify the influence diagram to account for the competitive risk considerations.

To simplify the visual complexity of the diagram, we can collapse the operational expenses into a single node called Annual OPEX Module. Of course we can refer to the prior diagram to explain what the underlying components contain.

Whether RoadRunner appears on the market or not is represented by the node RoadRunner Comes to Market. On the condition that RoadRunner does successfully bring a product to market, they will do so over some period of time, reflected by the node Time Frame RoadRunner Comes to Market. Notice that in our diagram (Figure 3-7), now, we communicate our understanding that an uncertainty conditionally affects another uncertainty. This is also true of the Available Market Share, as it is conditionally related to the difference between RoadRunner's time frame of development and the construction duration of Phase 1. Finally, we also indicate that the Price Discount will be conditionally related to RoadRunner Comes to Market.

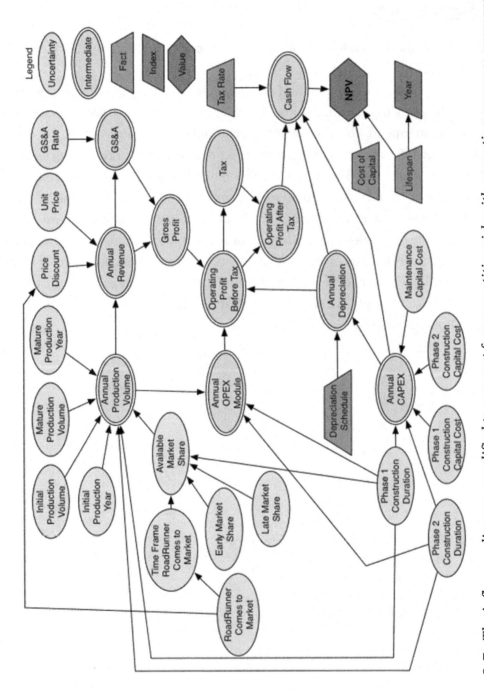

Figure 3-7. *The influence diagram modified to account for competitive risk with operating expenses aggregated into one intermediate node called the Annual OPEX Module*

Include the Run Index

Accommodating the risk layer of the model requires modifying our deterministic model in some important ways. First, we need to recognize that the single point p50 assumptions should be expanded to include an SME's belief about the likely range of each variable through the entire simulation. Remember that the three columns p10, p50, and p90 are the cumulative probability quantiles representing the SMEs' 80th percentile prediction interval for the outcome of our assumptions. Recall, too, that we will use the lines from the CSV data file (lines 15–19 in Figure 3-8) that contain the assumptions that reflect the competitive risk considerations.

	A	B	C	D	E	F
	variable	p10	p50	p90	units	notes
1	variable	p10	p50	p90	units	notes
2	p1.capex	35000000	40000000	47000000	$	phase 1 capital spend to be spread over the duration of Phase 1.
3	p1.dur	1	2	4	yrs	phase 1 duration.
4	p2.capex	15000000	20000000	30000000	$	phase 2 capital spend to be spread over the duration of Phase 1.
5	p2.dur	1	2	4	yrs	phase 2 duration.
6	maint.capex	1500000	2000000	3000000	$	annual maintenance spend.
7	fixed.prod.cost	2250000	3000000	4500000	$	annual fixed production costs.
8	prod.cost.escal	2.25	3	4.5	%/yr	the rate at which the fixed costs grow.
9	var.prod.cost	2.625	3.5	5.25	$/lb	the variable production costs.
10	var.cost.redux	3.75	5	7.5	%/yr	the rate at which the variable production costs decay.
11	gsa.rate	19	20	22	% revenue	general, sales & administration cost.
12	time.to.peak.sales	3.75	5	7.5	yrs	the number of years to reach max sales.
13	mkt.demand	3.75	5	7.5	ktons/yr	the amount of annual sales at peak.
14	price	4.5	6	9	$/lb	product price.
15	rr.comes.to.market	0	0.6	1		the probability that RoadRunner comes to market.
16	rr.time.to.market	4	5	6	yrs	the time elapsed until RoadRunner comes to market.
17	early.market.share	45	50	65	%	the market share available to Acme if RoadRunner comes to market early.
18	late.market.share	70	75	85	%	the market share available to Acme if RoadRunner comes to market late.
19	price.redux	12	15	20	%	reduction in price that that occurs if RoadRunner comes to market.

Figure 3-8. *The data table used for the deterministic business case in* `.csv` *format rendered in a spreadsheet*

The header lines of our new uncertainty-enabled R model (`Risk_Model.R`) look like this now:

```
# Read source data and function files. Modify the path names to match your
# directory structure and file names.
source("/Applications/R/RProjects/BizSimWithR/data/global_assumptions.R")
d.data <- read.csv("/Applications/R/RProjects/BizSimWithR/data/risk_
assumptions.csv")
source("/Applications/R/RProjects/BizSimWithR/libraries/My_Functions.R")

# Slice the values from data frame d.data.
d.vals <- d.data[, 2:4]
```

Next we define a probability distribution for the variables that will change with each iteration of the simulation. For this example, I used the BrownJohnson distribution described earlier. Notice, too, that I limited the range of the distributions to truncate at natural boundaries using the pmin() and pmax() functions as necessary. For example, because the Phase 1 capital cannot be negative, I surrounded the p1.capex definition with pmax(..., 0) to keep any stray simulation samples from going below 0. For definitions that need to be constrained between 0 and 1, I nested the pmin() and pmax() functions, as in

```
early.market.share <- pmin( pmax(..., 0) 100).
```

```
# Assign values to variables using appropriate distributions.
p1.capex <- CalcBrownJohnson(0, d.vals[1, 1], d.vals[1, 2],
    d.vals[1, 3], , kSampsize)
p1.dur <- round(CalcBrownJohnson(1, d.vals[2, 1], d.vals[2, 2],
    d.vals[2, 3], , kSampsize), 0)
p2.capex <- CalcBrownJohnson(0, d.vals[3, 1], d.vals[3, 2],
    d.vals[3, 3], , kSampsize)
p2.dur <- round(CalcBrownJohnson(1, d.vals[4, 1], d.vals[4, 2],
    d.vals[4, 3], , kSampsize), 0)
maint.capex <- CalcBrownJohnson(0, d.vals[5, 1], d.vals[5, 2],
    d.vals[5, 3], , kSampsize)
fixed.prod.cost <- CalcBrownJohnson(0, d.vals[6, 1], d.vals[6, 2],
    d.vals[6, 3], , kSampsize)
prod.cost.escal <- CalcBrownJohnson( , d.vals[7, 1], d.vals[7, 2],
    d.vals[7, 3], , kSampsize)
var.prod.cost <- CalcBrownJohnson(0, d.vals[8, 1], d.vals[8, 2],
    d.vals[8, 3], , kSampsize)
var.cost.redux <- CalcBrownJohnson( , d.vals[9, 1], d.vals[9, 2],
    d.vals[9, 3], , kSampsize)
gsa.rate <- CalcBrownJohnson(0, d.vals[10, 1], d.vals[10, 2],
    d.vals[10, 3], 100, kSampsize)
time.to.peak.sales <- round(CalcBrownJohnson(1, d.vals[11, 1],
    d.vals[11, 2], d.vals[11, 3], ,kSampsize), 0)
```

```
mkt.demand <- CalcBrownJohnson(0, d.vals[12, 1], d.vals[12, 2],
   d.vals[12, 3], ,kSampsize)
price <- CalcBrownJohnson(0, d.vals[13, 1], d.vals[13, 2],
   d.vals[13, 3], ,kSampsize)
rr.comes.to.market <- rbinom(kSampsize, 1, d.vals[14, 2])
rr.time.to.market <- round(CalcBrownJohnson(1, d.vals[15, 1],
   d.vals[15, 2], d.vals[15, 3], ,kSampsize))
early.market.share <- CalcBrownJohnson(0, d.vals[16, 1],
   d.vals[16, 2], d.vals[16, 3], 100, kSampsize)
late.market.share <- CalcBrownJohnson(0, d.vals[17, 1],
   d.vals[17, 2], d.vals[17, 3], 100,kSampsize)
price.redux <- CalcBrownJohnson(0, d.vals[18, 1], d.vals[18, 2],
   d.vals[18, 3], ,kSampsize)
```

The first few samples from p1.capex look like this:

```
[1] 43959855 56002835 35500058 36685885 40846491 44401842 40763622 38461247
38454413
[10] 39002984 37501273 37931937 43875385 36893128 29870863 36169428
40210438 44370703
```

After you have set the variables in your own code file, load the source into the R console and run a few of the variables to get a feel for how their results appear and make sure that all results conform to the distributions' definitions. Anomalies will indicate that you have possibly misdefined a variable. Of course, they can also indicate that your definition, although working just fine, is producing results that do not conform to the context of the business case. Instead of repairing the definition, you will likely need to rethink the definition employed or reassess the inputs.

We need to make one exception to using the BrownJohnson distribution for rr.comes.to.market, which represents the probability that RoadRunner comes to market. Although a broad discussion about the nature of probabilities is beyond the scope of this document, I maintain here without discussion that probabilities are not expressed as distributions. Only the uncertainty about event outcomes or measurable conditions are expressed as distributions. Because an event either occurs or not, it makes sense to describe a binary event with a single probability.[6] The appropriate distribution for this is the Bernoulli distribution, which is a special limiting case of the binomial distribution. So, for `rr.comes.to.market`, we can define it as `rbinom(kSampsize, 1, d.vals[14, 2])`, where the `prob` parameter is satisfied with the corresponding probability value in the CSV data file.

You will notice, however, that I placed a value of 0 for the `rr.comes.to.market` `p10`, and 1 for the `p90`. This is not to indicate an uncertainty about the probability in question. Rather, I use these values here to serve as placeholders to compare the effects in later sensitivity analysis of the value of no competition (`rr.comes.to.market` = 0) to being certain that competition will occur (`rr.comes.to.market` = 1).

Now, each variable defined earlier will simulate a series of outcomes. The length of each series is `kSampsize`. You can think of these series as existing in the `run` index. To run our simulation model, we want our R code to select a single value from each variable starting at `run` = 1, step through all the calculations in the same order as the deterministic model in the last chapter, and then repeat the process through each step in the `run` index until our simulation reaches `run` = `kSampsize`. Essentially, we want the sample values in the `run` index to operate orthogonally to the `year` index.

This now presents a little problem, though, because of R's vectorizing and recycling nature, which is usually a very powerful characteristic of the language. We naturally want to think of the `run` index as being orthogonal to the `year` index, but unless we take some steps to force that condition, R will recycle the values in the `run` index of each variable through the `year` index, completely messing up the indexing of the resulting calculations. As you might have guessed, we can use the `sapply()` function to address this issue.

To run the simulation, we wrap each variable calculation from our original deterministic model with the `sapply()` function, starting at the CAPEX module to the NPV, in which arrays of different shapes are applied to each other. Also, because `sapply()` wants to reorient the resulting calculations such that the `year` index becomes

---

[6]Describing a probability with a distribution naturally leads to the question of why we wouldn't then describe each probability within that distribution with another distribution, ultimately leading to an infinite regress. Instead, we think about probabilities as degrees of belief that well-defined events will occur or not. The actual probability assigned to a specific event results from a ratio of bets one would place for and against the event happening. Because one would only have a fixed purse of money to bet, the final allocation of bets are single points, not uncertain ranges. In short, probabilities are, themselves, statements of uncertainty. We don't need to compound uncertainty further by layering on more uncertainty. Oops, I said I wouldn't get into this discussion here.

rows and the run index becomes columns, we transpose each calculation with the t() function to maintain the calculations in the run x year orientation so that we can easily read results of intermediate calculations in the R console when the run index is longer than the year index, as will most likely always be the case.

```
# CAPEX module
phase <- t(sapply(run, function(r) (year <= p1.dur[r]) * 1 +
  (year > p1.dur[r] & year <= (p1.dur[r] + p2.dur[r])) * 2 +
  (year > (p1.dur[r] + p2.dur[r])) *3))

capex <- t(sapply(run, function(r) (phase[r, ] == 1) * p1.capex[r]/
p1.dur[r] +
  (phase[r, ] == 2) * p2.capex[r]/p2.dur[r] +
  (phase[r, ] == 3) * maint.capex[r]))

# Depreciation module
depr.matrix <- array(sapply(run, function(r) sapply(year, function(y)
ifelse(y <= p1.dur[r] & year>0, 0,
  ifelse(y == (p1.dur[r]+1) & year<y+kDeprPer & year>=y, p1.capex[r] /
  kDeprPer,
    ifelse((year >= y) & (year < (y + kDeprPer)), capex[r, y - 1] /
    kDeprPer, 0)
    )
  )
)), dim=c(kHorizon, kHorizon, kSampsize))

depr <- t(sapply(run, function(r) sapply(year, function(y)
  sum(depr.matrix[y, , r])))))

# Competition module
market.share <- (rr.comes.to.market ==1) * ((rr.time.to.market <= p1.dur) *
  early.market.share/100 + (rr.time.to.market > p1.dur) *
    late.market.share/100 ) +
  (rr.comes.to.market == 0)*1

# Sales module
mkt.adoption <- t(sapply(run, function(r) market.share[r] *
  pmin(cumsum(phase[r, ] > 1) / time.to.peak.sales[r], 1)))
```

```r
sales <- t(sapply(run, function(r) mkt.adoption[r, ] * mkt.demand[r] *
  1000 * 2000))
revenue <- t(sapply(run, function(r) sales[r, ] * price[r] *
  (1 - rr.comes.to.market[r] * price.redux[r]/100)))

# OPEX module
fixed.cost <- t(sapply(run, function(r) (phase[r, ] > 1) * fixed.prod.
cost[r] *
  (1 + prod.cost.escal[r]/100)^(year - p1.dur[r] -1)))
var.cost <- t(sapply(run, function(r) var.prod.cost[r] *
  (1 - var.cost.redux[r]/100)^(year - p1.dur[r] -1 ) * sales[r, ]))
gsa <- t(sapply(run, function(r) (gsa.rate[r]/100) * revenue[r, ]))
opex <- fixed.cost + var.cost

# Value
gross.profit <- revenue - gsa
op.profit.before.tax <- gross.profit - opex - depr
tax <- op.profit.before.tax * kTaxRate/100
op.profit.after.tax <- op.profit.before.tax - tax
cash.flow <- op.profit.after.tax + depr - capex
cum.cash.flow <- t(sapply(run, function(r) cumsum(cash.flow[r, ])))

# Following the convention for when payments are counted as occurring
# at the end of a time period.
discount.factors <- 1/(1 + kDiscountRate/100)^year
discounted.cash.flow <- t(sapply(run, function(r) cash.flow[r, ] *
  discount.factors))
npv <- sapply(run, function(r) sum(discounted.cash.flow[r, ]))
mean.npv <- mean(npv)
```

As we discussed earlier in the development of the depreciation matrix in Chapter 2, the sapply() function effectively iterates across an index just like a for loop, except it does so in one line of code. For example with phase, as R steps through the run index, a sample value from each of p1.dur and p2.dur is selected out of their respective distribution arrays, and the logic performed against the year index is as if the deterministic model were in play. When the calculation is complete, the resultant value is placed in the rth row of the phase array. The logic is similar for capex and all the other intermediate calculations through mean.npv. You can follow the entire R code for the risk model in Appendix B.

Once we reach NPV simulation, we can easily find the mean NPV:

```
mean.npv <- mean(npv)
```

No other function is required but the `mean()` function because all indexes but the `run` index have been reduced out, leaving only the samples in `run`. The `mean()` function simply finds the average of all the values in a given vector. The result gives us a depressing -$5,793,909. If we made decisions based on average values alone, we would stop here and go no further with our analysis. However, Chapter 4 will show us that there might yet be a reason to consider how to improve the value of this opportunity to our economic benefit.

When I calculate the mean NPV, I get -$5,793,909 on this particular run of the model. Running the model again, I get -$5,451,535. You will notice this same effect as well because of the simulation error inherent in simple Monte Carlo routines. One way to overcome this error, other than employing more advanced simulation routines (e.g., median Latin hypercube sampling), is to use more samples. We defined the sample size in our `global_assumptions.R` file as `kSampsize <- 1000`. Just change this value to a larger number to observe that the predicted mean NPV converges to a more stable position over several runs. On my computer, 1,000 iterations requires 1.542 seconds. I determined this value using the `system.time()` function.

```
system.time({source("/Applications/R/RProjects/BizSimWithR/Risk_Model.R")})
```

Changing `kSampsize` to 5,000 requires 7.307 seconds. The desire for improved precision comes at the price of a slightly longer runtime.

Just to illustrate how powerful the vectorizing `sapply()` function is compared to running the risk model in one big `for` loop, a similar model constructed in a `for` loop requires 2.701 seconds using a sample size of 1,000—an increase of 1.75-fold. The simulation using 5,000 samples requires 31.237 seconds, a greater than fourfold increase over the model using the `sapply()` function, using the same number of samples. Of course, the exact numbers you get will vary by your computing configuration (i.e., operating system, processor speed, RAM), but the relative differences between the vector versus the looping approach should be similar.

Moving along, we can follow the same `sapply()` logic in the construction of the pro forma statement based on the mean values of the pro forma elements.

```
# Pro forma
# Create a data frame of the variables' mean values to be used in the pro
# forma.
```

```r
pro.forma.vars <- array(
  c(
    sales,
    revenue,
    -gsa,
    gross.profit,-fixed.cost,
    -var.cost,
    -opex,
    -depr,
    op.profit.before.tax,
    -tax,
    op.profit.after.tax,
    depr,
    -capex,
    cash.flow
  ),
  dim = c(kSampsize, kHorizon, 14)
)

# Finds the annual mean of each pro forma element.
mean.pro.forma.vars <- array(0, c(14, kHorizon))

for (p in 1:14) {
  mean.pro.forma.vars[p,] <- sapply(year, function(y)
    mean(pro.forma.vars[, y, p]))
}

pro.forma <- data.frame(mean.pro.forma.vars)

# Assign text names to a vector. These will be the column headers of
# the data frame.
pro.forma.headers <-
  c(
    "Sales [lbs]",
    "Revenue",
    "GS&A",
    "Gross Profit",
```

```
    "Fixed Cost",
    "Variable Cost",
    "OPEX",
    "-Depreciation",
    "Operating Profit Before Tax",
    "Tax",
    "Operating Profit After Tax",
    "+Depreciation",
    "CAPEX",
    "Cash Flow"
  )

# Coerces the default column headers to be the headers we like.
colnames(pro.forma) <- year
rownames(pro.forma) <- pro.forma.headers
```

We will discuss the preparation for graphing of the cash flow and cumulative cash flow with confidence bands using the sapply() function in the next chapter.

CHAPTER 4

Interpreting and Communicating Insights

Certain graphical display devices help us to interpret and communicate powerful insights from the immense information produced by the Monte Carlo simulation process. These display devices visually communicate the range of exposure we face and the central tendency of outcomes we care about. They prioritize our attention on the risk factors that can affect us most.

Cash Flow and Cumulative Cash Flow with Probability Bands

To plot the cash flow and cumulative cash flow with their 80th percentile probability bands, we first specify a vector of the desired quantiles.

```
q80 <- c(0.1, 0.5, 0.9)
```

We then apply the quantiles() function in each year of the cash flow (Figure 4-1) and cumulative cash flow (Figure 4-2) using the sapply() function.

```
cash.flow.q80 <- sapply(year, function(y) quantile(cash.flow[, y], q80))
cum.cash.flow.q80 <- sapply(year, function(y) quantile(cum.cash.flow[, y],
q80))
```

© Robert D. Brown III 2018
R. D. Brown III, *Business Case Analysis with R*, https://doi.org/10.1007/978-1-4842-3495-2_4

To plot these results, use the following:

```
# Plots the 80th percentile cash flow quantiles.
plot(
  0,
  type = "n",
  xlim = c(1, kHorizon),
  ylim = c(min(cash.flow.q80) / 1000,
           max(cash.flow.q80) / 1000),
  xlab = "Year",
  ylab = "[$000]",
  main = "Cash Flow",
  tck = 1
)
lines(
  year,
  cash.flow.q80[1,] / 1000,
  type = "b",
  lty = 1,
  col = "blue",
  pch = 16
)
lines(
  year,
  cash.flow.q80[2, ] / 1000,
  type = "b",
  lty = 1,
  col = "red",
  pch = 18
)
lines(
  year,
  cash.flow.q80[3, ] / 1000,
  type = "b",
```

```
  lty = 1,
  col = "darkgreen",
  pch = 16
)
legend(
  "topleft",
  legend = q80,
  bg = "grey",
  pch = c(16, 18, 16),
  col = c("blue", "red", "dark green")
)
```

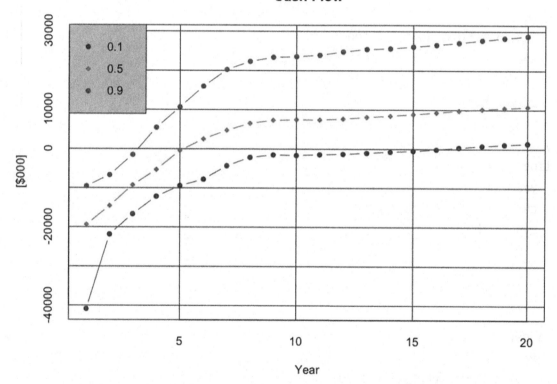

Figure 4-1. *The cash flow with probability bands around the 80th percentile prediction interval, representing the range of possible cash flows given all the relevant information we have at hand*

```r
 # Plots the 80th percentile cumulative cash flow quantiles.
plot(
  0,
  type = "n",
  xlim = c(1, kHorizon),
  ylim = c(min(cum.cash.flow.q80) / 1000,
           max(cum.cash.flow.q80) / 1000),
  xlab = "Year",
  ylab = "[$000]",
  main = "Cumulative Cash Flow",
  tck = 1
)
lines(
  year,
  cum.cash.flow.q80[1,] / 1000,
  type = "b",
  lty = 1,
  col = "blue",
  pch = 16
)
lines(
  year,
  cum.cash.flow.q80[2,] / 1000,
  type = "b",
  lty = 1,
  col = "red",
  pch = 18
)
lines(
  year,
  cum.cash.flow.q80[3,] / 1000,
  type = "b",
  lty = 1,
  col = "darkgreen",
  pch = 16
)
```

```
legend(
  "topleft",
  legend = q80,
  bg = "grey",
  pch = c(16, 18, 16),
  col = c("blue", "red", "dark green")
)
```

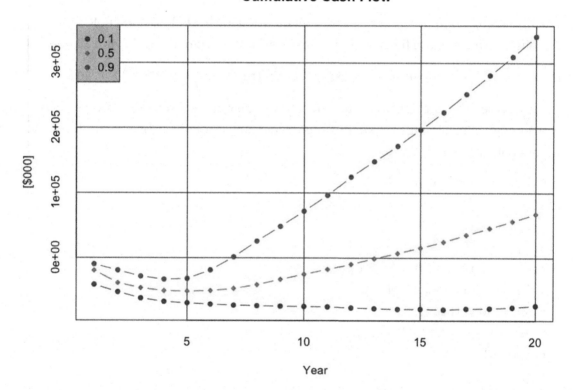

Figure 4-2. *The cumulative cash flow with probability bands, useful for determining the potential range of time in which payback on a cash basis can occur*

The cash flow graph (Figure 4-1) shows us that there is an 80% probability, given the quality of our information, that we can break even between the third and sixteenth year of operation, and the 50–50 outcome occurs around Year 5. The cumulative cash flow graph (Figure 4-2) is much less encouraging. It shows us there is an 80% probability that we will reach payback between Year 7 and never! The cash flow graph is important for showing when positive cash flow will likely occur, but the cumulative cash flow graph presents an initial insight into the more sobering reality of the economic value of the investment opportunity; that is, when an investment's resultant cash flows pay back the initial cash outlays. Of course, this result does not yet include the effects of the time value of money, so sobering results can often turn frightening when the discount rate is properly applied.

The Histogram of NPV

The histogram of the NPV is quite easy to set up using the hist() function of R. First, establish the locations of the histogram bins' breakpoints. Here we begin with 20 bins.

```
breakpoints <- seq(min(npv), max(npv), abs(min(npv) - max(npv)) / 20)
```

R sets these bin breakpoints automatically, but they are not frequently represented with enough detail. Using this parameter gives us some flexibility to play around with the number of bins and their size.

```
# Calculates and plots the histogram of NPV.
hist(
  npv / 1000,
  freq = FALSE,
  breaks = breakpoints / 1000,
  main = "Histogram of NPV",
  xlab = "NPV [$000]",
  ylab = "Probability Density",
  col = "blue"
)
```

Also, the hist() function plots the actual counts of the values that fall into the histogram bins. By setting freq = FALSE, we force the hist() function to use the probability density instead. The end result is demonstrated in Figure 4-3.

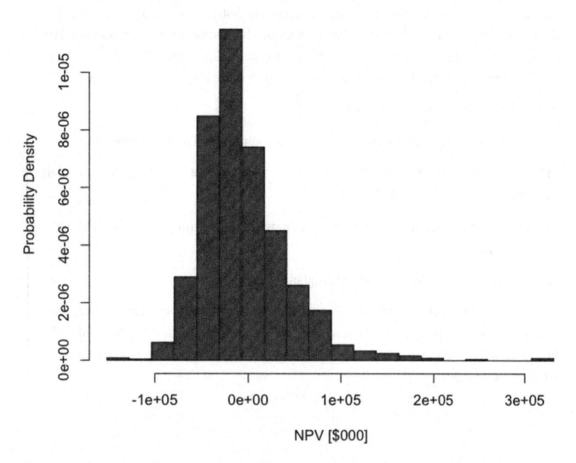

Figure 4-3. *The histogram of the NPV shows the central tendency of the net value of the investment opportunity*

The uncertainty in this problem has clearly produced a situation where the great bulk of the NPV outcomes will most likely fall below $0. In fact, on average they will fall below $0, as the mean NPV = -$5.8 million. Now we need to understand by just how much.

The Cumulative Probability Distribution of NPV

The approach to plotting the cumulative probability distribution is similar to that of the cash flow and cumulative cash flow confidence bands; however, we won't need to employ the sapply(), but we will need a longer vector of quantile values to get enough detail in the plot. We define the quantile breakpoint values by

```
cum.quantiles <- seq(0, 1, by = 0.05)
```

which gives a vector of 21 values from 0 to 1 with step sizes of 0.05, as shown here:

[1] 0.00 0.05 0.10 0.15 0.20 0.25 0.30 0.35 0.40 0.45 0.50 0.55 0.60 0.65 0.70 0.75 0.80 0.85 0.90 0.95 1.00.

Now, using the quantile() function again, we find the cumulative probability of the NPV at the 21 points.

```
cum.npv.vals <- quantile(npv, cum.quantiles)
```

If we put these values into a data frame, like cum.npv.frame <- data.frame(cum.npv.vals), we see in Figure 4-4 that we face an 80% probability that the NPV will fall between -$56.2 million and $56.6 million. Our probability for betting that the outcome will be less than $0 is ~62%.

	cum.npv.vals
0%	-126917826
5%	-67472787
10%	-56194253
15%	-48028127
20%	-41847105
25%	-37048318
30%	-32724226
35%	-29285101
40%	-25300330
45%	-20349067
50%	-14788718
55%	-10071990
60%	-4846563
65%	1122237
70%	7703073
75%	18029095
80%	28489400
85%	40675034
90%	56628131
95%	81272469
100%	395778771

Figure 4-4. *The domain and range of the cumulative probability of the NPV. The 80th percentile prediction interval, indicated in blue, shows there is ~62% probability that the outcome will be $0 or less.*

Now, to plot the NPV cumulative probabilities, as shown in Figure 4-5, we use the following code.

```
# Plot the cumulative probability NPV curve.
plot(
  cum.npv.vals / 1000,
  cum.quantiles,
  main = "Cumulative Probability of NPV",
  xlab = "NPV [$000]",
  ylab = "Cumulative Probability",
  "b",
  tck = 1,
  col = "blue",
  pch = 16
)
```

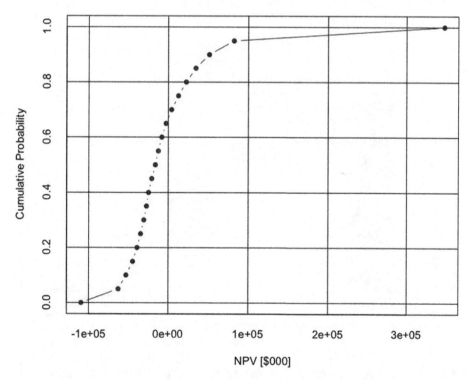

Figure 4-5. *The cumulative probability distribution of the NPV*

Therefore, by using both the probability density and the cumulative probability plots, we can observe where the bulk of the results would likely fall, as well as other important probability intervals (e.g., the 80th percentile prediction interval). We also see that the deterministic approach to the problem could have led us into trouble if we had stopped there—an expected loss of $5.8 million is certainly more sobering than the originally determined $8.2 million. Recall, though, that I emphasized that the initial deterministic results were tentative. We only acquire an informative picture of the decisions we make after we consider the effects of uncertainty and risk.

This is not the end of the analysis, though. There very clearly exists some opportunity to do better than less than $0. The question is this: What can we do to manipulate future outcomes to our benefit? Before we can answer that question, though, we need to know what we most likely should manipulate.

The Waterfall Chart of the Pro Forma Present Values

A waterfall chart, like the one shown in Figure 4-6, is useful for showing the relative cumulative contribution that the present value of the revenue (or accrued benefits) and cost elements make to the NPV.

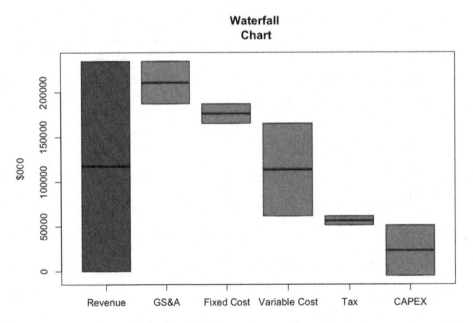

Figure 4-6. *The pro forma waterfall chart shows how the present value of the major pro forma line items contribute cumulatively to the NPV*

To create a waterfall chart, we first need to extract the appropriate benefit and cost elements from the pro forma array.

```
# Extract the rows from the pro forma for the waterfall chart.
waterfall.rows <- c(2, 3, 5, 6, 10, 13)
waterfall.headers <- pro.forma.headers[waterfall.rows]
wf.pro.forma <- pro.forma[waterfall.rows, ]
```

Next, we find the present value of these elements in the same way we found the NPV of the cash flow.

```
# Find the present value of the extracted pro forma elements.
pv.wf.pro.forma <- rep(0, length(waterfall.rows))
for (w in 1:length(waterfall.rows)) {
 pv.wf.pro.forma[w] <- sum(wf.pro.forma[w, ] * discount.factors)
}
```

We create a cumulative sum of the present values.

```
cum.pv.wf.pro.forma1 <- cumsum(pv.wf.pro.forma)
```

We duplicate the prior vector, but shift all the values forward one space, dropping the last value.

```
cum.pv.wf.pro.forma2 <- c(0, cum.pv.wf.pro.forma1[1:(length(waterfall.
rows) - 1)])
```

Now, to plot the floating columns of the waterfall chart, we need two vectors, one to capture the high position of each bar, and the other to capture the low position of each bar. We can do this by finding the pairwise maximum of the last two vectors, then by finding the pairwise minimum of the last two vectors.

```
wf.high <- pmax(cum.pv.wf.pro.forma1, cum.pv.wf.pro.forma2)
wf.low <- pmin(cum.pv.wf.pro.forma1, cum.pv.wf.pro.forma2)
```

Finally, we assign these last two vectors to an array.

```
waterfall <- array(0, c(2, length(waterfall.rows)))
waterfall[1, ]<- wf.high
waterfall[2, ]<- wf.low
colnames(waterfall) <- waterfall.headers
rownames(waterfall) <- c("high", "low")
```

We then plot the waterfall with the following code by co-opting the box plot.

```
# Plot the waterfall.
boxplot(
  waterfall / 1000,
  data = waterfall / 1000,
  notch = FALSE,
  main = "Waterfall Chart",
  xlab = waterfall.headers,
  ylab = "$000",
  col = c("blue", rep("red", 5))
)
```

If we want to improve profit by controlling costs, variable cost should get our first attention, followed by GS&A and CAPEX. However, don't assume that reducing these items arithmetically necessarily leads to increased profit, as there could be systematic effects that relate, say, capital spent now to operating costs incurred later. In a real analysis, this idea should be explored in a more detailed model for completeness, but for the purposes of this tutorial we won't consider this further.

The Tornado Sensitivity Chart

For our initial pass of analysis in Chapter 2, the deterministic sensitivity provides us some clues about which variables we should pay attention to, either in the form of manipulating them to our benefit or developing a mitigation plan to prevent some undesirable outcome in the objective function caused by them. However, this also assumes that the variables are under our control, that they are essentially decision variables. During the initial investigation and planning of a business opportunity, this condition of control rarely is the case.

As we've already recognized that the variables' actual future outcomes are most likely uncertain, in Chapter 3 we represented the variables in the model with distributions. Each distribution represents our conception of what the range of those outcomes could be with associated probabilities for intervals across the range. (This implies that a ±x% relative change in any variable might either be highly improbable or just a small fraction of the most likely range of actual behavior.)

What we need is a way to prioritize our attention on the uncertain variables, based on our current information about them and how strongly the likely range of their behavior might affect the average value of the objective function, NPV. So, before we go seeking ways to control variables on the guidance of the deterministic sensitivity analysis, we first need to understand if the probable range of outcomes for any variable is significant enough to matter to make a clear decision of any kind.

Tornado sensitivity analysis works by observing how much the average NPV changes in response to the 80th percentile range of each variable sequentially. We choose a variable and set it to its p10 value, then we record the effect on average NPV. Next we set the same variable to its p90 and record the effect on average NPV. During both of these iterations, we let the other variables run according to their defined distribution. Repeating this process for each variable, we observe how much each variable influences the objective function both by their functional strength and likelihood of occurrence.

We can make this observation easily with a floating bar chart, where each bar is assigned to a particular variable. The width of a bar runs from the low response of the average NPV to the high for the given variable. The bars are ordered such that the variable with the widest bar goes at the top and the narrowest is at the bottom. The declining widths of the bars give the chart its distinctive tornado, or funnel, shape.

To set up the code for this routine, we begin just as we did in the original deterministic sensitivity analysis by duplicating and modifying the base R code file (Risk_Model.R), then giving it a name like Risk_Model_Sensitivity.R. Our new file starts out looking just like the original.

```
# Read source data and function files. Modify the path names to match your
# directory structure and file names.
source("/Applications/R/RProjects/BizSimWithR/data/global_assumptions.R")
d.data <- read.csv("/Applications/R/RProjects/BizSimWithR/data/risk_
assumptions.csv")
source("/Applications/R/RProjects/BizSimWithR/libraries/My_Functions.R")

# Slice the values from data frame d.data.
d.vals <- d.data[, 2:4]
```

Now we add some additional lines to set up the process that will loop through our variables and create an initialized array to store the values of the NPV on each sensitivity iteration.

```
sens.range <- c(0.1, 0.9)
len.d.vals <- length(d.vals[, 1])
len.sens.range <- length(sens.range)
npv.sens <- array(0, c(len.d.vals, len.sens.range))
```

Just as before, we assign the appropriate distribution to each uncertain assumption and simulate them.

```
# Assign values to variables using appropriate distributions.
p1.capex <- CalcBrownJohnson(0, d.vals[1, 1], d.vals[1, 2],
  d.vals[1, 3], , kSampsize)
p1.dur <- round(CalcBrownJohnson(1, d.vals[2, 1], d.vals[2, 2],
  d.vals[2, 3], , kSampsize), 0)
p2.capex <- CalcBrownJohnson(0, d.vals[3, 1], d.vals[3, 2],
  d.vals[3, 3], , kSampsize)
p2.dur <- round(CalcBrownJohnson(1, d.vals[4, 1], d.vals[4, 2],
  d.vals[4, 3], , kSampsize), 0)
maint.capex <- CalcBrownJohnson(0, d.vals[5, 1], d.vals[5, 2],
  d.vals[5, 3], , kSampsize)
fixed.prod.cost <- CalcBrownJohnson(0, d.vals[6, 1], d.vals[6, 2],
  d.vals[6, 3], , kSampsize)
prod.cost.escal <- CalcBrownJohnson( , d.vals[7, 1], d.vals[7, 2],
  d.vals[7, 3], , kSampsize)
var.prod.cost <- CalcBrownJohnson(0, d.vals[8, 1], d.vals[8, 2],
  d.vals[8, 3], , kSampsize)
var.cost.redux <- CalcBrownJohnson( , d.vals[9, 1], d.vals[9, 2],
  d.vals[9, 3], , kSampsize)
gsa.rate <- CalcBrownJohnson(0, d.vals[10, 1], d.vals[10, 2],
  d.vals[10, 3], 100, kSampsize)
time.to.peak.sales <- round(CalcBrownJohnson(1, d.vals[11, 1],
  d.vals[11, 2], d.vals[11, 3], ,kSampsize), 0)
mkt.demand <- CalcBrownJohnson(0, d.vals[12, 1], d.vals[12, 2],
  d.vals[12, 3], ,kSampsize)
```

```
price <- CalcBrownJohnson(0, d.vals[13, 1], d.vals[13, 2],
  d.vals[13, 3], ,kSampsize)
rr.comes.to.market <- rbinom(kSampsize, 1, d.vals[14, 2])
rr.time.to.market <- round(CalcBrownJohnson(1, d.vals[15, 1],
  d.vals[15, 2], d.vals[15, 3], ,kSampsize), 0)
early.market.share <- CalcBrownJohnson(0, d.vals[16, 1],
  d.vals[16, 2], d.vals[16, 3], 100, kSampsize)
late.market.share <- CalcBrownJohnson(0, d.vals[17, 1],
  d.vals[17, 2], d.vals[17, 3], 100,kSampsize)
price.redux <- CalcBrownJohnson(0, d.vals[18, 1], d.vals[18, 2],
  d.vals[18, 3], ,kSampsize)
```

Next, we collect our simulated assumptions into an array that we iterate through as we sequentially replace each variable's simulated samples with the p10 and p90 quantile values.

```
d.vals.vect <- c(
  p1.capex,
  p1.dur,
  p2.capex,
  p2.dur,
  maint.capex,
  fixed.prod.cost,
  prod.cost.escal,
  var.prod.cost,
  var.cost.redux,
  gsa.rate,
  time.to.peak.sales,
  mkt.demand,
  price,
  rr.comes.to.market,
  rr.time.to.market,
  early.market.share,
  late.market.share,
  price.redux
)
d.vals.temp <- array(d.vals.vect, dim=c(kSampsize, len.d.vals))
```

The array `d.vals.temp` holds the original set of simulated samples. For the run of the sensitivity analysis, we borrow values from this array and place them in a duplicate array.

```
d.vals.temp2 <- d.vals.temp
```

We also need an array that contains only the p10 and p90 assumption parameters. We extract those from the `d.vals` array with the following:

```
d.vals2 <- d.vals[, -2]
```

Recall that the `d.vals` array looks like this:

```
             p10       p50        p90
 1   3.500e+07  4.0e+07  4.70e+07
 2   1.000e+00  2.0e+00  4.00e+00
 3   1.500e+07  2.0e+07  3.00e+07
 4   1.000e+00  2.0e+00  4.00e+00
 5   1.500e+06  2.0e+06  3.00e+06
 6   2.250e+06  3.0e+06  4.50e+06
 7   2.250e+00  3.0e+00  4.50e+00
 8   2.625e+00  3.5e+00  5.25e+00
 9   3.750e+00  5.0e+00  7.50e+00
10   1.900e+01  2.0e+01  2.20e+01
11   3.750e+00  5.0e+00  7.50e+00
12   3.750e+00  5.0e+00  7.50e+00
13   4.500e+00  6.0e+00  9.00e+00
14   0.000e+00  6.0e-01  1.00e+00
15   4.000e+00  5.0e+00  6.00e+00
16   4.500e+01  5.0e+01  6.50e+01
17   7.000e+01  7.5e+01  8.50e+01
18   1.200e+01  1.5e+01  2.00e+01
```

The subscript indicator `[, -2]` in `d.vals` tells R to remove the second column from `d.vals`, which is the p50 column.

Now, to summarize again, our R code will loop through our list of borrowed simulated values associated with each variable. On the selection of a specific variable that is being tested in the loop, the simulated samples in the variable's row will be replaced sequentially with its p10 value from `d.vals2`, and then its p90 value. Once the code has run its two tests at the sensitivity points on a given variable, the R code will restore the variable's original samples from the `d.vals.temp` array to the `d.vals.temp2` array and move on to the next variable to repeat the process.

Before we move on, let's think about the way we set up the deterministic sensitivity analysis. In that case, we stepped across the sensitivity points defined for the low, median (p50), and high values for our variables.

The central value of the sensitivity analysis was always the same because we were running a deterministic model. In stochastic systems, like the one we're representing in our uncertain model, the central value of the objective function (i.e., the mean) we find in the base uncertainty model will not necessarily be a function of the p50s of the assumptions' parameters. We therefore can't just find the center of the sensitivity analysis here by setting all the variables to their p50s in one step.

Another problem also arises from the simulation error we discussed in the last chapter. We likely will not get the exact same mean NPV from the samples simulated in the base uncertainty model as we would get from the samples simulated here in this instance of the sensitivity model, although they will probably be close for a large enough sample set. The way around this conundrum is to run the model twice with the same simulation samples, once to find the mean NPV, then once more to find the sensitivity of that mean NPV to the range in the assumptions. By doing this, we maintain the central value around which its sensitivity behaves. How do we do this? By placing our base model in one big function that has one parameter: the data array that contains the required samples for a given run of the model. We start the function at the point in our base uncertainty model where we extracted the values from our original data parameter array, simulated their samples with the appropriate distribution, and then assigned them to meaningful variable names. Our function won't need to resimulate the samples for each variable, though. It will simply use the samples we've already generated and placed in an indexed array. The following code represents the base model converted to a function that returns the samples for the NPV calculation.

```
CalcBizSim = function(x) {
# x is the data array that contains the presimulated samples for
# each variable.

        p1.capex <- x[, 1]
        p1.dur <- x[, 2]
        p2.capex <- x[, 3]
        p2.dur <- x[, 4]
        maint.capex <- x[, 5]
        fixed.prod.cost <- x[, 6]
```

```
        prod.cost.escal <- x[, 7]
        var.prod.cost <- x[, 8]
        var.cost.redux <- x[, 9]
        gsa.rate <- x[, 10]
        time.to.peak.sales <- x[, 11]
        mkt.demand <- x[, 12]
        price <- x[, 13]
        rr.comes.to.market <- x[, 14]
        rr.time.to.market <- x[, 15]
        early.market.share <- x[, 16]
        late.market.share <- x[, 17]
        price.redux <- x[, 18]

# CAPEX module
phase <- t(sapply(run, function(r) (year <= p1.dur[r]) * 1 +
  (year > p1.dur[r] & year <= (p1.dur[r] + p2.dur[r])) * 2 +
  (year > (p1.dur[r] + p2.dur[r])) *3))

capex <- t(sapply(run, function(r) (phase[r, ] == 1) * p1.capex[r]/p1.dur[r] +
  (phase[r, ] == 2) * p2.capex[r]/p2.dur[r] +
  (phase[r, ] == 3) * maint.capex[r]))

# Depreciation module
depr.matrix <- array(sapply(run, function(r) sapply(year, function(y)
ifelse(y <= p1.dur[r] & year>0, 0,
  ifelse(y == (p1.dur[r]+1) & year<y+kDeprPer & year>=y, p1.capex[r]
  / kDeprPer,
    ifelse((year >= y) & (year < (y + kDeprPer)), capex[r, y - 1]
    / kDeprPer, 0)
    )
  )
)), dim=c(kHorizon, kHorizon, kSampsize))

depr <- t(sapply(run, function(r) sapply(year, function(y)
  sum(depr.matrix[y, , r]))))
```

```r
# Competition module
market.share <- (rr.comes.to.market ==1) * ((rr.time.to.market <= p1.dur) *
  early.market.share/100 + (rr.time.to.market > p1.dur) *
    late.market.share/100 ) +
  (rr.comes.to.market == 0)*1

# Sales module
mkt.adoption <- t(sapply(run, function(r) market.share[r] *
  pmin(cumsum(phase[r, ] > 1) / time.to.peak.sales[r], 1)))
sales <- t(sapply(run, function(r) mkt.adoption[r, ] * mkt.demand[r] *
  1000 * 2000))
revenue <- t(sapply(run, function(r) sales[r, ] * price[r] *
  (1 - rr.comes.to.market[r] * price.redux[r]/100)))

# OPEX module
fixed.cost <- t(sapply(run, function(r) (phase[r, ] > 1) * fixed.prod.
cost[r] *
  (1 + prod.cost.escal[r]/100)^(year - p1.dur[r] -1)))
var.cost <- t(sapply(run, function(r) var.prod.cost[r] *
  (1 - var.cost.redux[r]/100)^(year - p1.dur[r] -1 ) * sales[r, ]))
gsa <- t(sapply(run, function(r) (gsa.rate[r]/100) * revenue[r, ]))
opex <- fixed.cost + var.cost

# Value
gross.profit <- revenue - gsa
op.profit.before.tax <- gross.profit - opex - depr
tax <- op.profit.before.tax * kTaxRate/100
op.profit.after.tax <- op.profit.before.tax - tax
cash.flow <- op.profit.after.tax + depr - capex
cum.cash.flow <- t(sapply(run, function(r) cumsum(cash.flow[r, ])))

# Following the convention for when payments are counted as occurring
# at the end of a time period.
discount.factors <- 1/(1 + kDiscountRate/100)^year
discounted.cash.flow <- t(sapply(run, function(r) cash.flow[r, ] *
  discount.factors))
```

```
npv <- sapply(run, function(r) sum(discounted.cash.flow[r, ]))
return(npv)
}
```

I recommend placing this function code in a separate .R file and then importing it at the beginning of the sensitivity analysis file using the source() function in the same manner as we did with the global assumptions and other functions.

To calculate the base mean NPV, we run the CalcBizSim() once with the values we assigned to the d.vals.temp array.

```
base.mean.npv <- mean(CalcBizSim(d.vals.temp))
```

Then, to find the sensitivity of the base mean NPV to the range in our assumptions, we loop across the secondary array that contains our borrowed samples, replacing each variable's samples with a vector of the same length containing the values of the p10 and p90 values.

```
for (i in 1:len.d.vals) {
  for (k in 1:len.sens.range) {
# For a given variable, replace its samples with a vector containing
# each sensitivity endpoint.
    d.vals.temp2[, i] <-  rep(d.vals2[i, k], kSampsize)

# Calculate the mean NPV by calling the CalcBizSim() function.
      mean.npv <- mean(CalcBizSim(d.vals.temp2));

# Insert the resultant mean NPV into an array that catalogs the
# variation in the mean NPV by each variable's sensitivity points.
      npv.sens[i, k] <- mean.npv
    }

# Restore the current variable's last sensitivity point with its original
# simulated samples.
    d.vals.temp2[, i] <- d.vals.temp[, i]
}
```

The remainder of the code works just like the deterministic sensitivity analysis that set up the graphical results.

```
# Assign npv.sens to a data frame.
var.names <- d.data$variable
rownames(npv.sens) <- d.data$variable
colnames(npv.sens) <- sens.range

# Sets up the sensitivity array.
npv.sens.array <- array(0, c(len.d.vals, 2))
npv.sens.array[, 1] <- (npv.sens[, 1] - base.mean)
npv.sens.array[, 2] <- (npv.sens[, 2] - base.mean)
rownames(npv.sens.array) <- var.names
colnames(npv.sens.array) <- sens.range

# Calculates the rank order of the NPV sensitivity based on the
# absolute range caused by a given variable. The npv.sens.array
# is reindexed by this rank ordering for the bar plot.
npv.sens.rank <- order(abs(npv.sens.array[, 1] -
  npv.sens.array[, 2]), decreasing = FALSE)
ranked.npv.sens.array <- npv.sens.array[npv.sens.rank, ]
ranked.var.names <- var.names[npv.sens.rank]
rownames(ranked.npv.sens.array) <- ranked.var.names

# Plots the sensitivity array.
par(mai = c(1, 1.75, .5, .5))
barplot(
  t(ranked.npv.sens.array) / 1000,
  main = "NPV Sensitivity to
  Uncertainty Ranges",
  names.arg = ranked.var.names,
  col = "red",
  xlab = "NPV [$000]",
  beside = TRUE,
  horiz = TRUE,
  offset = base.mean / 1000,
  las = 1,
  space = c(-1, 1),
  cex.names = 1
)
```

We now obtain a chart (Figure 4-7) very much like our deterministic sensitivity chart, except this one is based on the variation to our mean NPV caused by all the sources of variation and risk (and the bars are colored red). The chart tells us that the top five variables could easily turn a bad situation worse, but they also tell us that if we could harness the drivers behind the variation in each variable, effectively converting each of these uncertainties into a decision, we could create an immensely valuable opportunity (well, immensely so relative to the negative mean NPV we currently see).

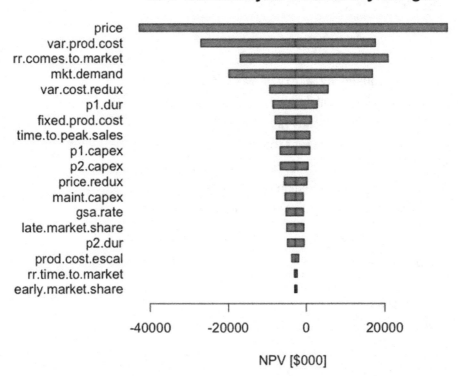

Figure 4-7. *The tornado chart displays the sensitivity of the mean NPV to the 80th percentile range of each uncertainty*

One of the important features of this type of sensitivity analysis is that it usually reveals that not all uncertainties are as important as we likely thought they were. In fact, we often find that at least one uncertainty is much more important than we anticipated, a feature of a good model.

Although it is beyond the scope of this book, the refinement in our analysis would be to find strategies that would potentially allow us to capture the upper end of each of those upper bars. The cumulative sum of the net value of reaching those goals would show us the marginal contribution of the value of control on those top four variables. After some market and competitive reconnaissance and creative planning, we might be able to find a way legally to deter RoadRunner from coming to market at all at little to no cost to us.

You can see the uninterrupted source code for this procedure in Appendix B.

Closing Comments

As I pointed out in Chapter 1, there are many ways the case study presented here could have reflected more complexity in the business environment, or the way the revenue and specific cost elements were represented could have taken completely different forms. My goal, though, in presenting this case study as a tutorial was to provide a platform for the following:

1. Thinking about R as a simulation alternative to spreadsheets.

2. Extending R from its common orientation of statistical analysis of empirical data toward an orientation more characterized as probabilistic reasoning.

The framework that we developed represented the mechanical structure for processing information we acquire through the reasoning process, but it by no means represents a complete analysis. We based our analysis on one design structure for engineering, market development, and operations. No one initial business solution can be thorough and comprehensive enough to address all the best (and worst) ways to allocate resources in the pursuit of success.

For our business cases analysis to be truly informative, we must consider multiple alternatives for achieving the same goal, simply because the process of value creation is pathway dependent. Each set of pathways faces opportunity costs that must be thoroughly explored before we can honestly say that we have performed our fiduciary responsibilities. In this tutorial, we considered only one pathway, one decision strategy. To truly learn what the future has to offer, we need to contrast and compare a good handful of thematically different ways we could take advantage of those potential offerings, measuring the risks and benefits of each, and discarding the ones that present the least opportunity to create value. We'll explore this alternative generation approach in Part 2.

This approach is the scientific process of thinking: We develop hypotheses about how we can create value in a given opportunity context, explore the conditioning effects for what imparts greater utility to those hypotheses, and discard the hypotheses that demonstrate weak explanatory power for creating value. This, in my opinion, is the best kind of process-driven, data-supported decision making.

PART 2

It's Your Move

Create valuable strategic decisions when you don't know what to do

CHAPTER 5

"What Should I Do?"

A few years ago, a friend of mine asked me to help him think through how to grow his business. He had successfully operated a small boutique professional services firm for several years up to that point. Now he was concerned about the demands on his personal time, his ability to save for retirement and his children's education, and a nagging sense that things were getting a little stale. Maybe he needed a new strategy, he thought. As we discussed his current situation, he expressed the following concerns:

- It's difficult to scale my company's growth.

- It's hard to find and keep the right people.

- My personal interests in creating a business are not yet realized.

- How can my company establish a means to generate ongoing revenue and equity value?

- How can my company make a much bigger and positive lasting effect on clients?

Once we clarified those concerns, we restated them as a strategic question.

How can we facilitate our clients' business development activity for maximum profit in such a way that our clients enjoy the way they create business, we track our clients' evolution of needs, and our services become a "must-have" habit for my company to grow a profitable, sustainable, and scalable business that I enjoy?

Considering the gravity of the strategic question, my friend wondered aloud, "But how do I get there? There are still so many decisions to make. I keep thinking about the pros and cons of each, but I don't quite know how to handle the complexity of all the possible combinations. What should I do?"

© Robert D. Brown III 2018
R. D. Brown III, *Business Case Analysis with R*, https://doi.org/10.1007/978-1-4842-3495-2_5

Usually, my clients initially face the very important issue of not really knowing what they want to achieve by making decisions, but in this case my friend knew his objectives fairly well. Where he was getting mired down was in conceiving effective ways to coordinate multiple decisions to achieve those objectives. Overwhelmed by the complexity of all the alternatives he faced, my friend didn't know what to do. Maybe you have found yourself in a similar situation.

Three Tools to Clarify Your Thoughts

When we make decisions, we usually speak of choosing among a known set of alternatives, a choice between a this or a that. We might think about some of these choices:

- Vanilla or chocolate or the new wasabi ice cream flavor.

- Pirelli tires or Michelin or Goodyear.

- A vacation in the Bahamas or one in Las Vegas.

If the alternatives are not mutually exclusive, we might even talk about choosing between some of this and some of that. These are the simpler decisions in life. Even if they are not routine, their consequences are usually limited in scope and magnitude. We usually address these decisions by weighing in our mind the net effect of the relative pros and cons of each choice without generally worrying that the consequences will be beyond our ability to handle should they turn out contrary to our preferences, even significantly so. If you do choose the wasabi-flavored ice cream and it's just awful (as I can assure you that it actually was), it's not an outcome you will regret the rest of your life. You'll gag a little and be a few dollars poorer—no big deal.

However, when we find ourselves facing complex personal or business situations, we are very rarely limited to the choices within the scope of a single decision category. Instead, we usually find ourselves thinking about the consequences of choosing among multiple coordinated alternatives across multiple decision categories. The number of potential decision alternatives we can choose from can lead to an overwhelming, swarming beehive of what-ifs in our minds.

For example, consider that you are a key decision maker within a company that develops biomedical devices. Choosing whether or not to launch a new product does not usually have a simple go or no-go set of alternatives. For a new product launch, you might have to think through the combined effects of choices about a new product's geographic distribution, pricing level, product configuration, packaging, staging, specific disease application, and so on. Each choice within a decision set leads to a set of implied potential consequences that, as a responsible decision maker, you should consider to the satisfaction of your fiduciary responsibilities. Although the associated economic analysis will represent enough complexity of its own, more immediately you might face some degree of confusion about which decision strategies you should even consider. You will probably not have time to consider all the possibilities.

Just among the decision categories listed in the biomedical device situation earlier, there exist hundreds, if not thousands, of possible decision strategies worth considering. What if each of the six categories contained three alternatives? The decision makers would face 3^6 combinations ($3 \times 3 \times 3 \times 3 \times 3 \times 3$) or 729 strategies in their analysis. What decision makers face is choice complexity, which is a significant contributor to two common decision failures: analysis paralysis and shooting from the hip.

Some decision makers forge ahead believing they should and can conquer the complexity by testing every possible scenario. However, as both the number of decision sets increase and the number of alternatives increase, the number of testable strategies can approach the number of atoms in the universe![1] Clearly, no one will ever commit to a decision while there is always one more scenario to test in this case. (I'm convinced that many people actually use analysis paralysis to avoid making decisions as a safe approach to managing their own career risk.) In the end, if anything ever gets done, it's usually too late to take advantage of the best part of the available value the opportunity provides.

Unfortunately, other decision makers opt to shoot from the hip as they assume that their gut is a better guide than the looming detailed analysis. Selective hindsight bias convinces them that their intuition is usually right (Isn't that what everyone thinks Malcolm Gladwell's book *Blink!*[2] taught us?). Or they reason that committing now and fixing things later (regardless of what the unforeseen costs might be) is at least movement, and movement is valuable for its own sake—until the bill for the unforeseen costs or unintended consequences comes due.

Is there a reasonable way to tame the choice complexity that you face as a decision maker in your real-world, complex decision situations? There is, using three simple

---

[1] I admit, this is a wee bit of an exaggeration, but as we'll see later in this tutorial, the number can be exceedingly large. Five to six orders of magnitude for the number of decision alternative combinations required to characterize a typical business problem is not uncommon.

[2] M. Gladwell, *Blink: The power of thinking without thinking* (New York: Little, Brown, 2005).

integrated thinking tools called a *decision hierarchy,* a *strategy table,* and a *strategy rationale table.* You should consider using these tools in these situations:

1. You experience a sense of being stuck about how to generate creative alternatives.

2. You lack a reasonable number of decision strategies with clearly understood descriptions that are not immediately recognized as worthy of consideration by key stakeholders, resulting in a general confusion about what to do.

3. You face a number of decision categories, each with two or more alternatives, such that you could reasonably combine the alternatives into a broad array of decision strategies. The strategies present the potential for a broad spectrum of resultant value.

Before you use these tools, you should be aware of their limitations. They will not tell you what to do or how to choose. Their only purpose is to help you create choices when you are stuck and feel as if you currently lack the creativity to do so. However, after some practice and repeated use, you will be surprised at how creative they allow you to be with the information you have at hand. You will come to see that the seeds of a solution usually lie within the definition of the problem itself. The three tools taken together will help you define and frame the problems or opportunities you face with greater clarity and creativity.

The Full Scope of Effective Decision Making

Alice came to a fork in the road. "Which road do I take?" she asked. "Where do you want to go?" responded the Cheshire Cat. "I don't know," Alice answered. "Then," said the Cat, "it doesn't matter."

—Lewis Carroll, *Alice in Wonderland*

Before I go further, I want to be clear about the value of this tutorial. It serves most effectively within a broader, more comprehensive framework for executive decision making. For such decision making to demonstrate the principles of a quality system, certain key questions need to be answered according to a process that refines and clarifies important information to those held accountable for making decisions prior to

CHAPTER 5   "WHAT SHOULD I DO?"

the commitment of resources that are intended to resolve the problem or capture the value associated with an opportunity.

The flowchart shown in Figure 5-1 depicts the process with many of the key questions associated with each phase of clarification and refinement. We understand the process to be a collaborative and open-ended effort within a commitment to make decisions in the best interests of the stakeholders who the decision makers have been entrusted to represent. When the leaders of an organization adopt and incorporate this quality process into their managerial decision-making activities, they can reasonably certify with specific warranties that they have met their fiduciary duty of care (Figure 5-2).[3]

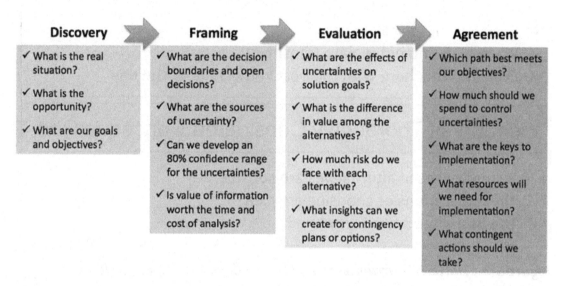

Figure 5-1. *At each phase of an effective decision management process, decision team members seek answers to key questions, resolve nagging issues, and synthesize important information about the current opportunity before making resource commitments to a course of action.*

[3]Ronald A. Howard, "Heathens, Heretics, and Cults." *Interfaces* 22(6), 1992, pp. 15-27.

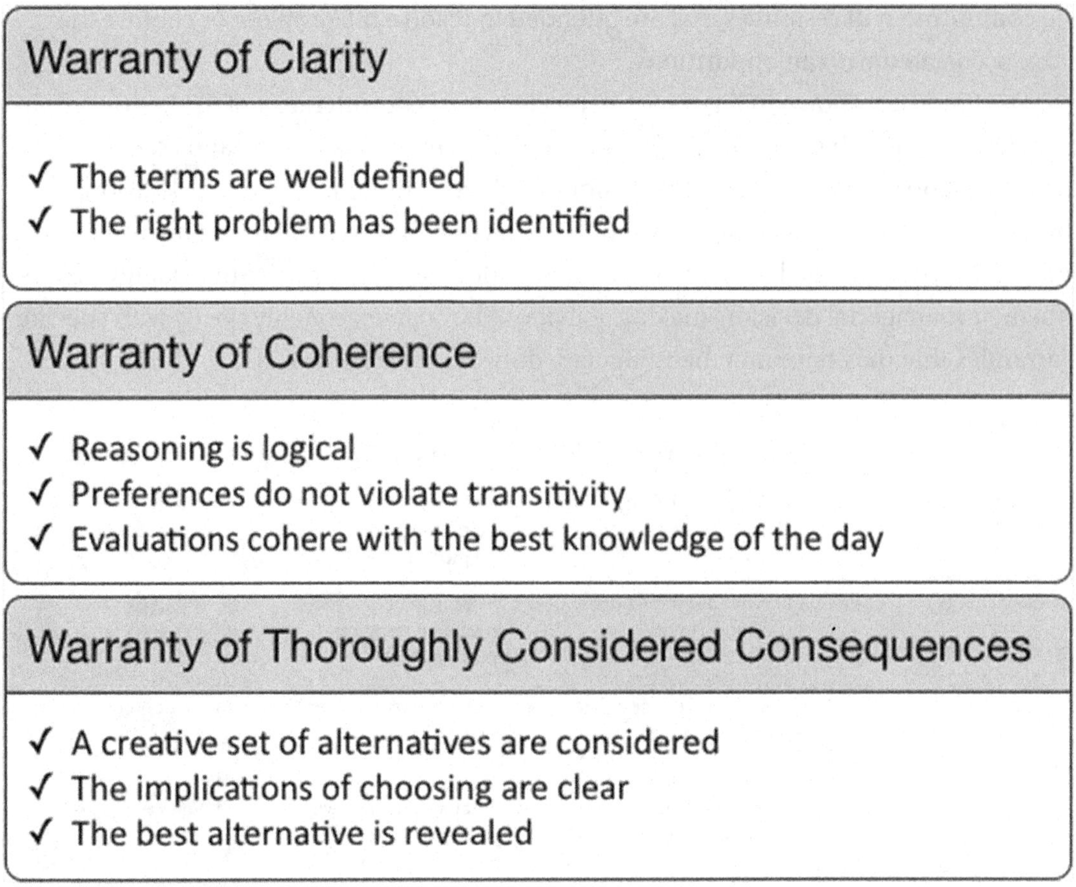

Warranty of Clarity

✓ The terms are well defined
✓ The right problem has been identified

Warranty of Coherence

✓ Reasoning is logical
✓ Preferences do not violate transitivity
✓ Evaluations cohere with the best knowledge of the day

Warranty of Thoroughly Considered Consequences

✓ A creative set of alternatives are considered
✓ The implications of choosing are clear
✓ The best alternative is revealed

Figure 5-2. *The three warranties provided by following an effective decision management process*

From the perspective of a full decision management process, this section will help you satisfy a small slice of the process and warranties just described:

- *Framing:* What are the decision boundaries and open decisions?

- *Warranty of thoroughly considered consequences:* A creative set of alternatives are considered.

Definitions

Throughout this tutorial, we use several terms that are colloquially interchangeable. To facilitate clear communication here, we use the following words with these stricter meanings.

1. *Decision:* An irrevocable allocation of a resource. A decision is not merely a mental disposition about what you might do. It represents an actual commitment to use of time, energy, money, or personnel.

2. *Decision type:* A classification for whether a decision is policy (foregone), strategic (open for consideration), or tactical (deferred). We discuss this more in the section on using the decision hierarchy.

3. *Decision hierarchy:* The set of decisions available for consideration in the context of the given problem or opportunity, each element of which has been assigned to its appropriate decision type.

4. *Strategic (or open) decision set:* A decision that is being considered in the current analysis that might contain several subchoices. The kind of dessert one might consider prior to dinner could be thought of as an open decision.

5. *Alternative:* One of the mutually exclusive choices available within a given strategic or open decision. Blueberry cobbler, apple cheesecake, or Boston cream pie might represent the available alternatives in the open dessert decision.

6. *Decision strategy:* A complex decision is one in which multiple open decision sets are considered in a coordinated manner under the effects of uncertainty. The set of selections of given alternatives within each of the open decisions under consideration represents a decision strategy. For example, thinking about a dinner decision strategy might include the following open decisions: drink, salad, meat, vegetable, dessert. The drink alternatives might be drink{water, wine, iced tea, Coca-Cola}. The alternatives in the other categories might be salad{Waldorf, carrot and raisin, iceberg wedge}; meat{beef, bbq pork, salmon}; vegetable{green beans, sweet potatoes, boiled radishes}; dessert{blueberry cobbler, apple cheesecake, Boston cream pie}. A decision strategy is based on the selection of one alternatives from each open decision. The effects of uncertainty on a given dinner decision strategy might be how much you will enjoy the combined tastes of the food after their

preparation, whether or not you will experience heartburn, or the
level to which you will increase your blood sugar or contribute to
your elevated cholesterol. Some people refer to a decision strategy
as a decision scenario.

What You Will Learn

The purpose of Part 2 is to give you the tools to dampen anxiety and clarify mental
confusion when you need to make numerous related decisions that might result in
complex interactions. By the concepts presented here, you will learn the following:

1. How to identify and partition important decision categories.

2. How to expand the space of decision alternatives associated with
   open decisions that require focused attention now.

3. How to create meaningful strategic themes based on the
   expanded set of decision alternatives.

4. How to manage and communicate the clarification and
   development of strategic themes with the three tools.

Complementary Resource: Integrated Decision Hierarchy and Strategy Table Template

To follow this discussion more effectively and to provide you with a template that you
can use in your own organization, I have provided the three tools in an integrated Excel
workbook that I developed and use in my own client engagements. You will be able to obtain
that file as part of the downloads available for this book at www.apress.com/9781484234945.
Before you read the rest of this chapter, I encourage you to download the integrated Excel
workbook that contains the decision hierarchy, strategy table, and strategy rationale table.
Save the downloaded file as a master copy and duplicate it as needed for each decision-
making exercise you encounter.

## CHAPTER 6

Use a Decision Hierarchy to Categorize Decision Types

When it comes to making decisions associated with a given opportunity, one mistaken assumption people frequently make is to think that they need immediate resolution on every decision—right now! Very rarely is this assumption true. We need a way to identify where our attention should be focused, however, and where we can defer our attention to avoid the inefficiency of that way of thinking.

Solving a problem usually involves breaking that problem down into useful categories of information. I call this process *partitioning*. Before we use the strategy table, we are going to partition some information so that we can identify actionable information and reduce the overall amount of information we have to deal with. The first tool we can use for this exercise is a *decision hierarchy*.

The first step is to conduct a brainstorming-like meeting with fellow members of the decision team. Tell the participants some time in advance (about a week) the purpose of the meeting and what type of information they should bring (i.e., ideas about decisions that need to be made in the course of addressing the problem under consideration). Emphasize that the meeting is not to decide what to do; rather, it is simply to consider what actions possibly need to be exercised without working the problem to an immature solution.

In the meeting, ask each participant to contribute one suggested decision issue. Record their decisions in the Decision Issues column in the Decision Hierarchy worksheet of the template.[1] Move to the next participant until everyone has had a chance to offer a suggestion. For now, do not allow any judgment about the applicability of a suggested

---

[1]Remember, you can download the template at `www.apress.com/9781484234945`. You don't necessarily need to use this template to gather the information used in this part of the decision framing process, but the template demonstrates good practices by putting everything in one place and it uses some code to help organize the flow and display of information.

© Robert D. Brown III 2018
R. D. Brown III, *Business Case Analysis with R*, https://doi.org/10.1007/978-1-4842-3495-2_6

decision to the current problem. Keep repeating this process until everyone has exhausted their suggestions. Letting each person offer just one suggestion at a time serves to prevent one person from dominating the meeting, and it allows people to strike off their redundant suggestions. The cyclic process might also spark new thoughts that people would not have had if they had just dumped their information at one time or in an isolated manner.

In the case of the friend who I introduced in Chapter 5, he and his partners identified the following decision issues:

- Should we offer a web service that enables clients?

- What size company should we aspire to be?

- Are there certain kinds of clients we won't take?

- Should we continue to offer our more public social and networking engagements (network gatherings, Chief Marketing Officer forums, online radio, etc.)?

- Should we deliver a commodity service or product?

- What kinds of business services should we offer?

- How deeply should we be involved in our clients' operations?

- What kinds of personnel should we hire?

- What is our focus choice of client?

- How should we price our services?

- What locations should we serve?

As your team considers the list of decisions, think about which ones might have already been decided for you as a matter of policy, legal, or regulatory constraint. Select Policy in the drop-down list to the right of the issue. These are decisions that you should consider foregone and won't need to actively consider as a value contributor. You will, however, need to consider them as you construct alternative combinations that we address in the next section. The policy decisions might function to require or preclude certain decision alternatives in strategic combination.

Once policy decisions are identified, discuss which of the remaining decisions really do need immediate attention or can be addressed after a commitment to a pathway is selected. Those decisions that need immediate attention should be classified as Strategic from the list of alternatives, and the remainder with the Tactical list alternative. This latter set, the Tactical decisions, are the deferred decisions that do not need any attention until later. If an active contention persists about whether a given decision should be considered strategic or tactical, go ahead and place it in the Strategic list. You can test the validity of that assignment as you go through the process of creating alternative combinations and, later, financial evaluation (see Figure 6-1).

	Decisions Issues	Type
1	Should we offer a web service that enables clients?	Strategic
2	What size company should we aspire to be?	Strategic
3	Are there certain kinds of clients we won't take?	Policy
4	Should we continue to offer our more public social and networking engagements (Network gatherings, CMO forums, online radio, etc.)?	Tactical
5	Should we deliver a commodity service/product?	Policy
6	What kinds of business services should we offer?	Strategic
7	How deeply should we be involved in out clients' operations?	Policy
8	What kinds of personnel should we hire?	Strategic
9	What is our focus choice of client?	Strategic
10	How should we price our services?	Strategic
11	What locations should we serve?	Strategic

Figure 6-1. *A raw list of decision issues classified by the decision team after thorough discussion about their relevance to the opportunity at hand*

Once you classify your decision issues to your satisfaction, sort the Decision Issues and Type array using Type as the sort column. Copy the Policy issues to the red policy cells, the Strategic issues to the green cells, and the Tactical issues to the yellow cells (see Figure 6-2). You might want to distill the issues into more succinct phrases.

Decision Hierarchy

Policy Decisions

(Requirements or constraints, unchangeable by this project team.)

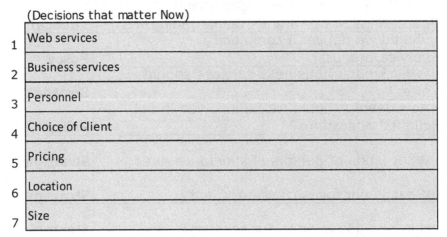

1	We won't compromise the quality of the deliverable
2	We won't deliver a commodity service/product
3	We will be highly selective with who we take as clients
4	We will be integrally involved in our clients' business to ensure results

Strategic (Focus) Decisions

(Decisions that matter Now)

1	Web services
2	Business services
3	Personnel
4	Choice of Client
5	Pricing
6	Location
7	Size

Tactical Decisions

(Decisions that can be deferred)

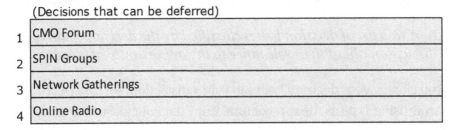

1	CMO Forum
2	SPIN Groups
3	Network Gatherings
4	Online Radio

Figure 6-2. *The decision hierarchy helps a decision team clarify and partition decision issues by whether an issue should be treated as foregone, open, or deferred. This allows the team to use its time most effectively by focusing its attention on the open issues that matter most.*

116

Once Strategic decisions have been settled on, begin to identify an exhaustive set of mutually exclusive alternatives that each decision set could represent to your organization. By exhaustive, I don't mean that you should include every possible alternative that anyone could exercise for a given decision category; rather, I mean only those choices that are meaningful to your organization in the context of the given opportunity.

Mutual exclusion can be a little more difficult to understand. In some cases, the alternatives might simply be yes and no. Choosing one would naturally exclude the other. In other cases, the decision alternatives might exist along a continuum. For example, suppose you are considering the acquisition of new combustion generators and you want to place that capital within the guidance of a green policy that exceeds the current regulatory requirements for the presence of a certain particulate in the generators' exhaust. You might define the alternatives as nonoverlapping bins within the considered range of particulate concentration, each bin corresponding to a level of capital commitment. Finally, some alternatives might represent combinations of alternatives. For example, the geographic scope alternatives for the delivery of a new product might originally include the United States, Europe, Asia, and the Middle East. Of course, there is no intrinsically logical reason you have to pick one over the others. You might, in fact, consider combined alternatives such as United States and Europe and Asia, or Middle East and Asia.

If you created a list of Strategic decisions in the decision hierarchy worksheet, you will notice that there are cells to the right of each decision with names that correspond to the decision name and an alternative number. Fill in the alternatives for each decision, writing over the existing formula in the template that generated the placeholder name, so that it looks like the worksheet in Figure 6-3. Although the template provides the space to have as many as 15 alternatives per decision, 15 is an exceedingly large number of alternatives for most strategic decisions.

Strategic (Focus) Decisions	Decision Options -->			
(Decisions that matter Now)	1	2	3	4
Web services	Business Opportunity Analysis	Targeting Tools	Communication Tools	Connecting Tools
Business services	Strategy	Capabilities	Execution	Strategy/Capabilities
Personnel	Contract resources	Alliance partners	Employees	Equity Members
Choice of Client	Professional Service Firms	General Corporations	Business to Professional	Challenger Brands
Pricing	Pay for Performance	Fixed Fee	Hourly	Service Guarantee
Location	Atlanta	Southeast	National	Global
Size	Sole Practitioner	Small Boutique (<25)	Big Firm	

Strategic (Focus) Decisions					
(Decisions that matter Now)	5	6	7	8	9
Web services	Capability Tools	Accountability & Tracking Tools	Content	Execution Enablement	None
Business services	Strategy/Execution	Capabilities/Execution	All		
Personnel	Contract & Alliance				
Choice of Client					
Pricing	Multi-year Contracts	Membership Fee			
Location	Virtual	Major Cities			
Size					

Figure 6-3. *Assign a set of actionable alternatives to each Strategic decision*

Note from Figure 6-3 above, which addresses my friend's business expansion question, he identified nine "Web Services" alternatives, and three business "Size" alternatives. Given all his identified alternatives, he faced $9 \times 7 \times 5 \times 4 \times 6 \times 6 \times 3 =$ 136,080 possible strategic combinations. The next section will show you how to tame this complexity into a much more manageable set of decision strategies.

Once you have appropriately categorized the decision types and assigned the Strategic decision alternatives, you can consider the decision hierarchy to be effectively complete. That does not mean, though, that the decision hierarchy cannot be amended. Executive decision making should be an open-ended, exploratory process that incorporates and synthesizes new and compelling information as it becomes available. If the hierarchy needs to be expanded or pruned as new information becomes available, then make sure you do so.

Before going further, though, the team should discuss the following questions:

- What problem are we really trying to solve and are these decisions appropriate?

- Are the policy decisions really givens or just team perceptions? Can they be changed? Who outside the decision team can resolve that ambiguity if it exists?

- Are all of the strategy or focus decisions included in the list?

- Are the decisions included appropriately defined?

- Are the decisions listed really decisions?

- Are the tactical decisions really dependent on the strategy decisions?

When you are finished with this step, you will again have reduced the decision complexity by effectively partitioning the potential decisions under consideration into tractable types with an understanding of their hierarchical relationship to each other.

## CHAPTER 7

Tame Decision Complexity by Creating a Strategy Table

A strategy is essentially a pathway for creating value or getting more of what we want with the least expenditure from what we already possess. The amount of value we create depends on the pathway we take in the process of executing a strategy. The strategy table is a tool for developing several creative pathways that could serve us in our attempt to get to our desired goals.

Probably one of the best ways to understand a strategy table is to consider a privateer who has obtained a secret map to a buried treasure of Spanish bullion and wants to develop a strategy to obtain it. Before the privateer begins marking out a pathway to achieve that goal, he first considers his current position, then he clarifies his objectives: make a net profit that minimizes risk to life and limb. As he looks at his map, he realizes there are many pathways that can be taken. In fact, technically speaking, there are an infinite number of pathways that can be taken. Some pathways might require him to traverse high mountains, territories with hostile residents, blazing deserts, fire swamps, raging rivers, seas with carnivorous shrieking eels, and towering cliffs. Depending on which route he takes, each based on a selection of subroutes, our privateer faces a set of costs and risks that differ by the route taken.

Even though there are an infinite number of pathways to consider, our privateer realizes that he doesn't need to think about all of them. The reason for this is that for any given pathway, other pathways that are close by probably don't differ that much in cost and risk. Instead, our privateer considers thematic threads across his map that are implied by the contours of the geography and cultural contexts of regions on the map. One thread might take the high road through mountainous passes and the raging rivers where the locals are friendly and generous, but the lower pathway through the fire swamp with its many pitfalls and predators would serve to deter his thieving competitors. The thread through the central desert presents the shortest pathway, but few make it across the desiccating sands before dying from dehydrating madness. Each thread represents a spanning set of global value, each of its subelements hanging together to support an implied theme instead of simply piecing together elements that independently seem to represent some locally optimal choice. It is usually not the case that the sum of subset optima produce a global system optimum. The strategy table helps us to avoid thinking that a good strategy is just the combination of independently good choices.

Now, back to our decision team. With the open decisions and their alternatives now presented to the decision team, the team should consider how one alternative from each open decision can build a fundamental theme. I recommend starting with what many of my colleagues refer to as the *momentum* strategy. This is the strategy or pathway you are already on (sometimes erroneously referred to as "do nothing") or considering or might already be developing. Consider how you are currently predisposed to allocate resources and develop that idea. Make sure you only select combinations that are reasonable or logical in their combination. On the strategy table in the strategy table worksheet, name the current strategy in a cell in the Strategic Theme column. Then, replicate the colored object in the cell and place a copy in a desired alternative cell underneath the names of the Strategic decisions that have been transposed from the decision hierarchy worksheet.

Now, develop more strategic combinations that are thematically different from the momentum strategy. After you develop the first two or three, stretch your imagination and think of the most aggressive or extreme strategy possible, or even one that the team thinks of as dumb. You might not ever actually implement either of these latter strategies, but considering them will provide you with at least one of the following benefits.

1. You will confront biases toward actions that can be taken by your organization and possibly find surprising wisdom in alternatives that had been too easily dismissed in the past.

2. You will preemptively put to rest any future suggestion to implement that strategy by those who aren't as close to the current opportunity context, and you will know why it likely won't work, with the ability to communicate those reasons objectively.

Figure 7-1 shows how my friend thought about linking decision alternatives together to produce a rich set of strategic themes.

Strategy Table

Focus Decisions -->

Strategic Theme	Web services	Business services	Personnel	Choice of Client	Pricing	Location	Size
Momentum "Strategic Coach"	Business Opportunity Analysis	Strategy	Contract resources	Professional Service Firms	Pay for Performance	Atlanta	Sole Practitioner
Virtual CRO	Targeting Tools	Capabilities	Alliance partners	General Corporations	Fixed Fee	Southeast	Small Boutique (<25)
Whole Enchilada	Communication Tools	Execution	Employees	Business to Professional	Hourly	National	Big Firm
Pursuit Planners	Connecting Tools	Strategy/Capabilities	Equity Members	Challenger Brands	Service Guarantee	Global	
Demand Gen. Agency.	Capability Tools	Strategy/Execution	Contract & Alliance		Multi-year Contracts	Virtual	
	Accountability & Tracking Tools	Capabilities/Execution			Membership Fee	Major Cities	
	Content	All					
	Execution Enablement						
	None						

Figure 7-1. *A completed strategy table in which each strategic theme is created by selecting and combining appropriate alternatives from each of the vertical strategic focus decision categories that cohere in some meaningful sense*

Use a specific icon to link the alternatives together. Once you have considered what the essence of a given combination represents, give it a memorable name that succinctly communicates that essence. In short, the strategy table functions as a working palette for the decision team to consider how the open, focus decision alternatives might be combined to create a valuable pathway to their objectives.

Be careful not just to create worst case, most likely, and best case scenarios and label them that way. You won't know what their likelihoods of success or failure are prior to quantitative evaluation anyway. Also, by conceiving and naming them in that manner, you will be advocating for the pathway you prefer (with all of its attendant biases) as opposed to exploring objectively which alternatives really do create value.

As we discussed earlier, one of the purposes of this thinking device is to avoid evaluating every possible decision alternative combination; however, as a rule of thumb, you should try to use every alternative at least once. In some cases, though, you will notice that even after thorough consideration, you choose only one of the alternatives for a given Strategic decision, or in other cases, you never choose one or more of the alternatives. Both cases reveal some interesting insights.

In the case of choosing just one alternative, you have discovered a heretofore implicit policy. In other words, the team has revealed that under the given circumstances it will always choose this one alternative. The less extreme situation of not choosing one or a few alternatives reveals that not all of the decision alternatives actually work in combination with the others. In my friend's case, his team decided that staying focused only in Atlanta was not an alternative they ever wanted to pursue. This observation leads to an extension of my recommendation in Chapter 6 regarding the lack of necessity of achieving immediate resolution on every decision; that is, you actually won't or can't consider every possible combination simply because some don't make sense together. As a result, again, your decision complexity is reduced.

As a final clarifying step, select the indicated alternatives from the drop-down lists in the appropriate cells for each strategic theme in the table below the working strategy table. Figure 7-2 shows what a simplified table looks like.

Strategy Table							
Focus Decisions -->							
Strategic Theme	**Web services**	**Business services**	**Personnel**	**Choice of Client**	**Pricing**	**Location**	**Size**
Momentum "Strategic Coach"	Content	All	Contract resources	Professional Service Firms	Fixed Fee	Major Cities	Sole Practitioner
Virtual CRO	Execution Enablement	Execution	Contract & Alliance	Professional Service Firms	Hourly & Service Guarantee	Southeast	Small Boutique (<25)
Whole Enchilada	Accountability & Tracking Tools, Execution Enablement	Execution	Equity Members	Challenger Brands	Multi-year Contracts	Virtual	Small Boutique (<25)
Pursuit Planners	Targeting Tools	Strategy/Execution	Alliance partners	Challenger Brands	Membership Fee	National	Small Boutique (<25)
Demand Gen. Agency.	Communication Tools	Capabilities/Execution	Employees	Challenger Brands	Pay for Performance	Global	Big Firm

Figure 7-2. *The simplified strategy table reduces the visual complexity of the working strategy table so that the essence of each strategy is easily communicated to others who might not be actively participating on the decision team*

Recall that when you began thinking about the decisions you needed to make in a complex business opportunity, you likely first wrestled with thinking through the pros and cons of each alternative independently. Then you realized that all the alternatives would need to coordinate in some way; therefore, all the coordinated combinations would have a set of benefits and risks beyond the independent value of each alternative taken alone. To make matters worse, the number of such combinations exceeded the number of decision alternatives with which you started. If there was any needle of value to be found in this haystack, the chances of finding it became even more remote. The strategy table tames that complexity and creates value simply by creating a map through the branching maze of alternatives. Alternative richness emerges from the fact that you've bounded the initial chaos of alternative combinations and reduced the ambiguity about meaningful and actionable ways to combine them. Initially facing paralysis by complexity, you now have reasonable strategic solutions to consider.

## CHAPTER 8

Clearly Communicate the Intentions of Decision Strategies

The strategy table requires a good deal of thought to create, but to an outsider to the current decision process the results will look simply like a mechanical combination of alternatives. The strategy table does not communicate well the meaning and context of each decision strategy or record the thinking that went into its construction. This is the point at which we provide that context so that the decision team members can continue to discuss the strategies in a consistent manner and so that others who are not centrally involved in the planning process (yet might have some peripheral, yet important, contribution) can appreciate the direction of our thought process. We provide the context and definitions in a qualitative description or strategy rationale for each decision strategy.

Just as is the case with the strategy table, the qualitative descriptions do not tell us what to choose or how to choose it. They are simply a device that provides a succinct description of the strategies being considered with enough requisite detail so that when we discuss these strategies with members of the decision team or others, referring to the name of the strategy is enough to have a meaningful discussion without a tedious recollection of exhaustive details. The practice among my colleagues is to provide these descriptions on one page per strategy in the following form:

1. *Name:* This is simply the name of the strategy.

2. *Description:* Provide a short rationale for why this strategy is being considered and what its key theme addresses.

3. *Objectives:* Describe the compelling goals and objectives that might be obtained.

© Robert D. Brown III 2018
R. D. Brown III, *Business Case Analysis with R*, https://doi.org/10.1007/978-1-4842-3495-2_8

4. *Benefits:* Name the most compelling benefits that you might obtain by taking this strategy.

5. *Risks:* Name the most compelling undesirable outcomes you might experience by taking this strategy.

6. *Wins if:* What intermediate events must come together to make this strategy succeed?

7. *Loses if:* What intermediate events could come together to make this strategy fail?

8. *Hunch:* How does the decision team think this strategy will play out if actually implemented?

An example is shown in Figure 8-1. You can use this information to compare with your final decision to see how biased or insightful your thinking might have been originally.

Strategy Name: (e.g., Momentum)	
Description:	
Objectives:	
Risks:	**Benefits:**
Loses if:	**Wins if:**
Hunch:	

Figure 8-1. *The table for capturing the qualitative description for a given decision strategy*

Once a qualitative description has been filled out for each decision strategy, the decision team should reflect on the descriptions and consider if any of them should be discarded. The decision to do this should be based on a recognition that the decision strategy does not cohere with the overall intention and purposes of the company or represents a pathway that simply fails to satisfy the objectives of the decision opportunity.

Be careful: Discarding a strategy shouldn't occur too quickly out of hand, be based on personal preferences, or be based on a hunch about economic value. Let the actual financial analysis discredit a strategy if financial concerns are the motivation. In too many cases in the past I've worked with clients who wanted to discard a strategy because they thought it would be too capitally aggressive or expose them to a potentially high probability of economic failure only to discover that the strategy was competitively viable after appropriate financial evaluation. Discard a strategy only on the basis of thematic weakness at this point. If you can discard a strategy, you have further reduced the decision complexity you face.

The qualitative description table in Figure 8-2 contains the rationale for the momentum strategy for my friend's business growth. We also developed rationales for each of the remaining four strategies we developed.

Strategy Theme: Momentum – Strategic Coach

Description: This is about Content and Coaching. Personal, individualized strategy, coaching and implementation of business development efforts for professional services firms in major U.S. cities. Online activities are limited to marketing and free thought leadership content distribution and some coaching. Provide one-stop shop for growth services through our alliance program. Contractors only besides founder.	
Objectives: Capture greater share of client marketing budget by delivering broader range of services through alliance; Remain lean and flexible; Emphasize content/thought leadership.	
Risks: ○ Coaching moves out of favor & commoditized ○ Competitors launch a scalable growth model ○ Content and coaching not effective enough online ○ Difficulty in keeping contractors engaged ○ Alliance wastes time, delivers very little ○ Founder still doing most of the heavy execution and little time is left to develop and market the content. ○ Sell 1x1 at individual professional level – low revenue/client, high churn, highly local	**Benefits:** ○ Little additional investment required ○ Alliance helps broaden service offerings ○ Fits founder's ability to generate content ○ Low personnel commitment ○ Might travel less through online coaching
Wins if: Not making additional investment in people, promotion or product/technology enables more near term profit/cash to pay down loans and fund lifestyle and moderate additional growth. **Loses if:** Not making additional investment in people, promotion or product/technology causes us to be outpaced and outmoded by competition and too little time to adequately service clients and sell.	
Hunch: Insufficient. Change or die.	

Figure 8-2. *A completed qualitative description table for a decision strategy should succinctly capture the essence of a strategy on one page*

Before moving forward with financial analysis of the decision strategies, consider the following questions:

- Who was involved in developing these alternatives?

- Was the right mix of disciplines, functions, and departments consulted in the development?

- Are the alternatives unique and compelling?

- If we are making market-facing decisions, what alternatives are we considering relative to our competitors and customers?

- Are there any alternatives that would be difficult for our competitors to replicate?

- Are there alternatives that customers would hate or find surprisingly delightful?

- Do the decision strategies represent a spectrum of mild to wild, or are they merely variations on a theme?

- What is the most extreme alternative developed? Does it really represent the potential to explore untested approaches to creating value that might have been dismissed in the past or never before considered by the organization?

With qualitative descriptions in hand, your team will now possess a set of thematically distinct and compelling decision strategies that by careful consideration have reduced the dizzying decision space you originally faced to a manageable set (approximately four to eight) that can be evaluated quantitatively. Where you started with confusion and ambiguity, you now have creative solution hypotheses that can be tractably compared and contrasted for opportunity cost analysis and decision trade-offs in a financial model.

## CHAPTER 9

What Comes Next

After a decision problem has been properly framed with well-designed decision strategies, it's time to evaluate the strategies on their potential ability to create value. You must not, however, use the wrong kind of evaluation method to avoid the more difficult effort of uncertainty analysis. Do so at your own peril.

What You Should Do

To reiterate, the purpose of this tutorial is to provide you with a mechanism to partition decision alternatives' complexity into a manageable level of detail that will permit you to conduct effective business case analysis that neither mires you in analysis paralysis nor gives short shrift to uncertainties and risks. If you follow the guidance outlined in this tutorial, when you face a complex decision situation, you will obtain a manageable set of decision strategies with their accompanying descriptions and qualitative assessments that put you in position for that next step of analysis. Conceivably, you might be able to choose clearly the decision strategy you should take at this point. It's conceivable, but not advisable. In my nearly 20 years of doing some form of decision and risk analysis and planning, only two or three times have I seen a client get to this stage knowing clearly what to do. The next step—which we explore in the first and fourth sections of this book—requires that we build a requisite, probabilistic financial model that compares and contrasts the relative potential value and risk of each decision strategy.

I briefly lay out the elements of that important set of next steps here:

1. Build an influence diagram that relates the decisions and conditional uncertainties to an objective, such as maximizing stakeholder value.

2. Identify internal SMEs who can explain why each uncertainty might vary as a function of a chosen decision strategy.

© Robert D. Brown III 2018
R. D. Brown III, *Business Case Analysis with R*, https://doi.org/10.1007/978-1-4842-3495-2_9

3. Build a quantitative, probabilistic model that incorporates the information supplied by the SMEs and gives you the ability to do the following:

    - Explore the systematic relationships between the decisions and the uncertainties on the objective.

    - Prioritize your attention on critical uncertainties that could potentially make you regret choosing one strategy over the others so that you can develop tactics to mitigate the risk they might lead to.

    - Understand and communicate the value of trade-offs for choosing one decision over the others; that is, thoroughly comprehend the opportunity costs.

What You Should Not Do

You should build a model to assess the value of your decision strategies before you commit to any one of them. What you should not do is build the wrong kind of model, and I mean a very specific kind of wrong kind of model. Usually, some decision maker will tell you that you don't have enough time or resources to build the right kind of model or populate it with the right kind of information. That decision maker is wrong and has usually not thought about how much time and additional expense will be incurred (which can be orders of magnitude larger than the cost of the planning phase in which mistakes are relatively cheap) to fix the problems caused by using the wrong kind of models to support complex business decision analysis. This business analog of technical debt in which a seemingly simple or easy solution is adopted instead of one that is more appropriately thorough yet takes a little more time to prepare has been the source of much weeping, wailing, and gnashing of teeth in the business world. The tendency to use these wrong kinds of models is usually accompanied by frustrated and impatient admonitions that "We have to do something now!"

Instead of the right kind of model, the—let's graciously assume for now—well-intentioned but misguided decision maker will suggest some kind of weighted objectives scoring model by which each decision strategy is rated on an ordinal scale (e.g., 1–5) as to how well it potentially satisfies a set of criteria that are each given their own importance weightings. You should run from these approaches as fast as you can. In all seriousness, research by Douglas Hubbard[1] and others has shown that these approaches appear to lead people to value-destroying decisions more often than not. In fact, it might even be the case that simply randomly selecting a decision strategy is more effective than using weighted objectives scoring types of selection models. Hubbard reported that people use these models, though, because they provide the emotional satisfaction of doing something simple that still appears to have some rigorous quantitative support behind it. In other words, they provide a placebo effect for the users.

However, apart from the simplistically satisfying emotional appeal of these types of models, there are several other reasons why they do not support good, effective decision making.

1. You will most likely fail to compensate mentally for the systematic variation in the key figures of merit that can be caused by underlying decisions, conditional uncertainties, and other factors. Instead, you will rely on estimating the value of key figures of merit at too high of a level of aggregated information. You might not be cutting down a tree with your bare hand, but you will be using a dull hatchet.

2. You will not understand the potential value and risk in dollars of the proposed decision strategies. Consequently, you will not know how to prioritize all the other opportunities and problems you might need or want to address in the most effective economic manner.

3. You will not know the rational upper bound on the amount you should spend to get better information (i.e., the value of information) about the uncertainties you face. In other words, your research budget will be another blind guess that compounds your exposure to risk.

4. You will not know how to blend the best aspects of the two best competing decision strategies into a superior hybrid strategy, nor will you know the value of doing so (i.e., the value of control).

---

[1]D. Hubbard, *The Failure of Risk Management: Why It's Broken and How to Fix It* (Hoboken, NJ: Wiley, 2009, pp. 237–238); D. Hubbard and D. Evans, "Problems with Scoring Methods and Ordinal Scales in Risk Assessment." *IBM Journal of Research and Development* 54(3), 2010, pp. 2:1–2:10.

In other words, if you're lucky, you might choose a satisfactory
solution when just a little additional effort could lead to a solution
that the stakeholders would value more. Not expending the
additional effort puts you in a position of potentially violating your
fiduciary responsibilities. Yikes!

Please don't misunderstand my point. I am not saying that if you do use these types
of models that you will never experience favorable outcomes. What I am saying is that if
you do use these types of models, you will possess little insight into why you experienced
the outcomes you did and what you could have done to lead to better outcomes in the
most economically effective manner.

Now, it's your move.

PART 3

Subject Matter Expert Elicitation Guide

Assess uncertainties when you don't have (much) data or a clairvoyant

CHAPTER 10

"What's Your Number, Pardner?"

Imagine that I need an important piece of information from you. It's important because the information will be used in a business case analysis in which millions of dollars will potentially be allocated to a given project. The allocation will occur if the indicative figure of merit exceeds zero. It's possible that the response you give me for the information I need could influence the decision to pursue the project or not, depending on the strength of the functional relationship of your information to the figure of merit. Likewise, assuming there is a strong relationship, the actual accrued value of the project could depend on the accuracy of your information in the following way, as illustrated in Figure 10-1:

1. You provide **accurate** information that corresponds to the state of actual accrued value of the project. When your information implies a profit, the project is a go, and the project actually captures value; or, when your information implies a loss, the project is a no-go, and the deferred project avoids losing value.

2. You provide **inaccurate** information that fails to correspond to the state of the actual accrued value of the project. When your information implies a profit, the project is a go, but the project actually loses value; or, when your information implies a loss, the project is a no-go, so the project fails to capture the value that was waiting for it.

© Robert D. Brown III 2018
R. D. Brown III, *Business Case Analysis with R*, https://doi.org/10.1007/978-1-4842-3495-2_10

		Information Response	
		Accurate	Inacccurate
Actual Accrued Value	Profit	Go --> Value Captured	No Go --> Value Deferred
	Loss	No Go --> Loss Avoided	Go --> Value Lost

Figure 10-1. Possible outcomes associated with the accuracy of information used to predict the value of a project endeavor

How do you feel? Does this feel momentous to you? Do you feel anxious that you'll be right?

"C'mon, I need a number from you. What is the highest point in Texas?" I urge you, revealing the momentous information that I need.

If you are not a proud citizen of the great state of Texas, the correct answer might elude you. You hesitate a bit. "Well, there are a lot of big pickup trucks and ten-gallon hats. Should I include those?"

"No, no," I reply. "I need the highest natural point in Texas."

"I recall that Houston seems to be right near the Gulf of Mexico, so that's sea level. But Dallas is flat as a fritter and, as I recall, not at a very high elevation. Maybe 400 feet above sea level. Of course, there's the Texas hill country, but I've never heard anyone talk about Texas mountain country. So, I'd say it's less than the definition of a mountain's height, which I think is about 2,000 feet. I'll say half that, 1,000 feet."

"Are you sure about that? You seemed to think through several mitigating conditions and caveats."

"No, I'm not sure, but to be safe, I'll say 1,000 feet, give or take 25%."

"I need a number. But I guess we can do a sensitivity test using 750 feet or 1,250 feet. That feels safe."

Of course, this exchange would be silly. Hopefully no one would base a business case on a guess when such information is easily accessible. The easiest way for you to answer this question would simply be to consult an expert on Texas geography or, if this question had been asked of you prior to the advent of the Internet, the most recently published geographical surveys in an encyclopedia or an official almanac might suffice. In this hyperconnected age of instant searches, the cheapest, most accessible source of information could be gleaned by a casual search of Google or Wikipedia. Facts such as these are easy to obtain these days. Truly, the prior scenario was laughable.

However, I suspect that you might have knowingly laughed because you actually have gone through a process like it, although not for numbers that are so easily verified.

Instead, you might have tried to establish values for events or outcomes that have not yet occurred to represent a key assumption for a given scenario in a business case analysis. Maybe these values are annual demand and revenue for a new product, or the capital cost required to produce a plant to satisfy product demand, or the staffing required to satisfy all the activities in a project work breakdown structure, or the duration of specific events in the work breakdown structure.

Whatever the case, you probably went through a process where you tried to recall important events or easily visualized memories of similar outcomes or related information. You anchored on this number, and then you thought about interacting forces or mitigating circumstances that might modify your anchor up or down. If your thinking was progressive enough, you might actually try to accommodate some wiggle room with a "reasonable" padding, like ±20%. An alternate approach you might have taken would be to assume the business case was either in best case, worst case, or most likely case scenarios, and then you tried to reason yourself into acceptable assumptions that would lead to the particular case.

What is the problem with these approaches if we don't have good data to use? The answers is two words: bias and arbitrariness.

The bias comes from the mental anchoring process that relies on easily available information. In fact, there is a very well-documented bias called the availability bias or heuristic[1] that catches people off guard quite frequently. Sometimes an attempt to adjust for being off just a bit includes the "reasonable" padding, but little thought is actually given to whether the range reflects what your measured uncertainty might be, meaning that you are willing to take a bet with stated odds that the actual value will fall between some low and high bracketing values. The range of padding is merely arbitrarily chosen because it feels like it accommodates some reasonable level of error. What you're left with is a biased anchor point with arbitrary error bounds.

In the other case in which you assume some scenario level of outcome (e.g., worst, most likely, best), the likelihood that all of the necessary assumptions would line up to produce any of the cases would be extremely low. In this case, the range of potential outcomes predicted by your analysis might be completely unrepresentative of the odds that you actually face.

You might be surprised to find out that the highest point in Texas is 8,749 feet above sea level. You might also be surprised to find out that I ask this same question

---

[1] https://www.behavioraleconomics.com/mini-encyclopedia-of-be/
availability-heuristic/

in seminars on decision and risk analysis, and I routinely get rationales from the participants that proceed like the dialogue presented earlier. The average response I receive has been around 1,000 feet, and the range of error around ±25%. Over nearly 20 years and dozens of seminars attended by many dozens of participants, I have observed no improvement in the ability of my highly educated and intelligent participants to think through this simple question that actually accounts for their uncertainty about the right answer (see Figure 10-2) in an unbiased manner.

Guadalupe Peak, also known as Signal Peak, is the highest point in Texas. It is in Guadalupe Mountains National Park three miles west of Pine Springs in northwestern Culberson County (at 31°53' N, 104°52' W). Its summit, with an elevation of 8,749 feet above sea level, rises 3,100 feet above Pine Springs.[2]

Figure 10-2. *The location of Guadalupe Peak in the Guadalupe Mountain Range in west Texas*

---

[2]Handbook of Texas Online, "Guadalupe Peak," http://www.tshaonline.org/handbook/online/articles/rjg19, 2018.

It's not as if I asked about the highest peak on Mars or some esoteric exoplanet orbiting the star Proxima Centauri. No, the question is about information that pertains to our own backyard, so to speak. Nor is it the case, as I just suggested, that the regular attendees are somehow mentally or educationally deficient. They are not. The problem is that all of us tend to approach the way we think about information that isn't fully certain to us in a systematically biased and arbitrary manner. Sure, most of us, when pressed about it, will admit that single point assumptions possess almost no possibility of being "right," but we usually aren't sure how we can improve the direction of the rightness. When we do try to admit some error, the arbitrary range we choose is really just an attempt to cover ourselves rather than a thorough exploration of factors that could lead to extreme outcomes. When we consider all the biased and arbitrary assumptions that are usually applied in project planning and business case analysis, we shouldn't be too surprised at the terrible failure rate of projects and business case outcomes we witness, a failure rate that could be dramatically improved if a more accurate assessment of uncertainties and risks were applied appropriately from the beginning (notwithstanding just pure bad luck and poor execution).

The standard answer today in the age of Big Data and data science is that we can avoid all those problems of bias and arbitrariness if we just use a lot of real, empirical data to test our assumptions. I don't necessarily disagree with that, as long as we have data that are consistent with the problem at hand, that are clean, that carry a clear provenance, and that are accompanied with explanations for why some data might have been excluded while the rest have been retained. If we have all that, that's great. The problem is that we often don't even get the luxury of having any data, let alone bad data, for planning and business case analysis. We might have some, but we often won't have all that we think we really need. The question we face, then, is: how we can construct some appropriate level of knowledge and information that controls for the bias of human minds when we don't have access to good data to support accurate planning and analysis?

The answer is, in some way, bleak. Unless we have access to an omniscient and unbiased source of information, like a benevolent clairvoyant, we won't be able to fully escape the effects of bias. The path forward will be to recognize that people with expertise, SMEs, are like a kind of database, a meat database that is subject to spoiling, but a database, nonetheless. They know stuff. They have some good information inside them, and they understand fine distinctions about the nature of the behavior and causes of events within the scope of their expertise. The goal will be to extract from them, to elicit, their specialized knowledge about pertinent events in such a way that controls for the biases–from which no one seems innately exempt–that cloud people's judgment at times.

What are the characteristics we should look for in a desirable SME, and is there a replicable, transparent process that allows us to control for any residual bias that might corrupt their expertise? The answer to both of those questions is the subject matter of this section.

What You Will Learn

The purposes of this section are to do the following:

1. Help you recognize appropriate SMEs to use in assessing information about uncertain events.

2. Provide you with a kind of script for helping SMEs acknowledge and assess their own bias.

3. Provide you with a routine to elicit the appropriate information from SMEs required to support high-quality business case analysis.

It will be incumbent on the reader to understand that the second and third points are not "turn-the-crank" methods for extracting perfectly precise information. Rather, they will be thinking devices that must accompany seasoned facilitation techniques that can only be improved with time and practice. Furthermore, the reader must recognize that if better information (i.e., more accurate, more precise, more affordable, and relevant) were available, then that information should be used.

The methods outlined herein are not supposed to provide a lazy substitute for good statistical analysis. Rather, the approach outlined here starts us down the path of acquiring and constructing mostly accurate information—a good prior, if you will—from SMEs. Once we submit this constructed information to the proper kind of sensitivity and value of information analysis, we will have the best prioritization of attention to update our current information with higher quality data and statistical analysis. Until we get to that point, we rely on data mining the minds of SMEs.

CHAPTER 11

Conducting SME Elicitations

SME elicitation is the process by which we alleviate, at least temporarily, SMEs from the effects of their biases so that they can think more clearly and creatively about what the future might hold or what might veil their ability to possess certain knowledge. During the process, we facilitate an enumeration of causative factors that can affect the outcome of an uncertainty that was identified in the influence diagram, and then we assign—via the expert guidance—a measurement to the uncertain outcome as a range of probabilities that spans the range of the outcome.

The Good SME

Unlike the rest of us, SMEs are especially versed in the underlying causes and mechanisms, along with the nuances and fine distinctions, that affect events associated with their area of concern, study, or responsibility. SMEs are not just scientists or industry gurus, although they certainly can be. They are people who possess close at hand familiarity with an idea, a process, a program, or an activity. They could be a lead researcher in your R&D group, or they could be the director of sales, an account manager, or the lead supervisor for a given shift. The consideration for who is an SME is not based so much on their credentials as it is their experience and deep working knowledge.

Furthermore, in contrast to a novice who goes through the thinking process we observed earlier to produce a single point (and falsely precise) prediction (Figure 11-1), a good SME explores his or her experience and comprehension of a subject to consider the reasons for why extreme outcomes of a particular event in his or her area of concern might occur. Once a given event has been defined well enough for an SME to consider, his or her response should be to tell you all the reasons for why a wide range of outcomes

© Robert D. Brown III 2018
R. D. Brown III, *Business Case Analysis with R*, https://doi.org/10.1007/978-1-4842-3495-2_11

can occur and what those outcomes might be. You will not get a single point estimate from a good SME. Instead, you will get a range, and not just a simple range, but a special kind of range called a *distribution.* A distribution will contain several values along the spectrum of the conceivable outcome with probabilities associated with those values. The probabilities will necessarily be logically coherent; that is, their sum along the range must strictly equal 1. These probabilities represent the subjective degrees of belief in the SME's mind about the likelihood that any particular interval within the range of concern will occur relative to the remaining intervals.

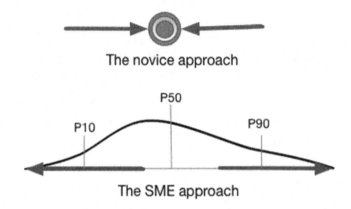

Figure 11-1. *The thinking process of a novice leads to posturing cleverness that results in a much too narrow consideration of outcomes compared with the range of outcomes proposed by an SME*

The novice essentially postures a level of precision that admits to requiring a higher quality of information than the current level of uncertainty about a situation reasonably permits. In this sense, they are producing an assumption, a statement about the state of nature in a given context that borders on axiomatic. SMEs, on the other hand, provide assessments; that is, conditional and contingent statements about the possibilities of outcomes that reflect their internal mental disposition that they just don't know precisely what will occur. SMEs are skeptics.

We can think about accuracy versus precision by visualizing several kinds of firearm patterns on a target produced by marksmen of varying degrees of competency, as shown in Figure 11-2. Novices lack control and do not generate reproducible patterns. Controlled bias leads to precision that scores low. Experts frame a bullseye accurately, and improving competency leads to a tighter pattern that continues to frame the bullseye.

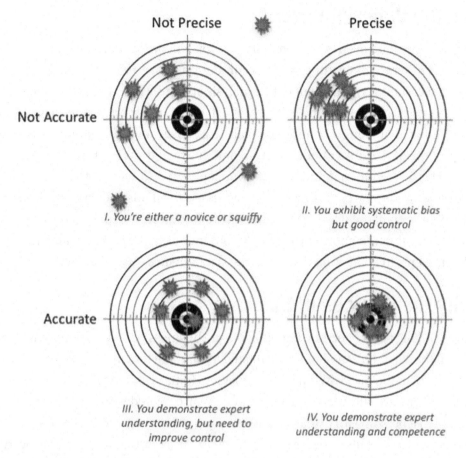

Not Precise   Precise

Not Accurate

I. You're either a novice or squiffy

II. You exhibit systematic bias but good control

Accurate

III. You demonstrate expert understanding, but need to improve control

IV. You demonstrate expert understanding and competence

Figure 11-2. *An example of accuracy versus precision*

Conduct the Assessment

The first goal of good SME elicitation is to obtain, in the absence of comprehensive, relevant data, an accurate representation of the natural distribution of an uncertain event over time or to characterize the possible range in which an event outcome might occur. An SME assessment incorporates the following steps:

1. Define the uncertain event.

2. Identify sources of bias.

3. Postulate and document the causes of extrema.

4. Measure the range of events with probabilities.

Define the Uncertain Event

Defining the uncertain event simply means that you will produce a statement that allows you or others to determine, without contention, that an event occurred or that it occurred in a certain way pertinent to the analysis at hand when the facts of the event are examined a posteriori. Determine if the event will occur in discrete values (e.g., [True or False], or [A, B, C]) or along a continuous scale (e.g., costs, revenue, temperature, mass). Explore how the SME thinks about the event and in which units (if any).

The goal is to produce a definition that would pass the so-called clarity test. The clarity test requires that a clairvoyant, who has access to all possible information, be given the definition of the uncertain event. The definition should permit the clairvoyant to make a determination about the event without requiring the clairvoyant to make any special judgments to determine that his or her data satisfies the definition.

Identify the Sources of Bias

Common thinking traps or errors that often plague business people are entitlement bias and wishful thinking. *Entitlement bias* is related to thinking that just because one has been successful in the past with important business initiatives that future success is not only possible, but virtually certain. This behavior is also related to *expert bias,* a state of mind in which a person with extensive and often publicly recognized experience in a field of practice regards his or her predictions about future outcomes as incontrovertible. *Wishful thinking* occurs when people believe that their plans will work as designed simply because they planned them and visualized the outcome ("If we build it, they will come"), or because they have strong emotional investment in the outcome that they desire.

Some of the common cognitive illusions that show up in uncertainty assessments are availability bias and anchoring. *Availability bias* is the tendency to use recently observed, emotionally impactful, or easily accessible information to lend belief that a certain state of affairs is the case or that an eventual outcome will be the case. Availability bias has a close cousin in *confirmation bias* (sometimes called "cherry picking"), which is the selective observation of evidence used to justify or buttress preconceived notions or closely held beliefs. To make a playful inversion on the warning commonly printed on rearview mirrors, availability bias drives the perception that images in the mirror are actually closer than they really are. *Anchoring* is the tendency of the mind to work from the first value that it conceives as a reference point for all other variation around it. Both availability bias and anchoring drive a tendency to believe that initial impressions serve as best guesses for most likely future outcomes.

To begin addressing these biases, ask the SME to describe any recent events or memorable events that come to mind associated with the event in question. Has something related to the event in question appeared in the news? Has there been a recent exposure to the event with a given value associated with the outcome?

If it is not obvious, ask the SME if there is a desirable outcome associated with the event. Would the SME, or those he or she supports, desire a higher or lower outcome or a targeted outcome?

Finally, explore how the outcome might affect the SME personally. Are there preferences, motivations, or incentives that might be associated with the SME and the event? Is the SME compensated in any way for the event occurring at some degree or level of outcome?

This line of questioning should help to reveal confirmation and availability bias, motivational bias, and wishful thinking, among others, at work in the mind of the SME. As a friend of mine says, "Confirmation bias is the root of all evil. And wishful thinking is root fertilizer." These must be rooted out and exposed.

Postulate and Document Causes of Extrema

Ask the SME to imagine the extreme case representing the opposite direction of the desired outcome. Explore what might have caused this outcome to occur. What had to occur for the future to be in the extreme outcome described? What are those factors? Repeat this process for the opposite extreme. Record the factors for each extreme in separate columns. When you finish tabulating the causative factors, place a quick rank of importance on the top three or four causes on the low and high side. Later, when you go back to consider risk mitigation on the uncertainty, if it's required, you might want to consider these higher ranked factors first for risk management or risk transfer.

Be sure to emphasize that at this point you are not seeking a number or a value. Resist the tendency to get a value as a starting point because that introduces anchoring. The objective is to counteract anchoring and availability biases by getting the expert away from their preconceived notions or influenced inclinations.

The simplified example shown in Table 11-1 represents how a construction engineer acting as an SME participated in the assessment for the duration of the structural steel construction for a chemical reactor. First, we simply discussed what was meant by "the duration of the structural steel construction." The SME defined this event as "the number of work days required to construct the mainframe steel, which included the building structure and pipe rack structure. It also included precast on the southern and eastern walls. The total duration would include the steel bid, fabrication, and erection." During our discussion, the engineer revealed that this particular part of the construction process on the kind of reactor being constructed contributed to the significant delay on another recent project, revealing that he had in mind an availability bias and might be inclined to provide a more pessimistic assessment about this case. Because his bias was for delay, we started thinking first about reasons why the shortest, most favorable duration was possible. The actual list was many elements long because the engineer understood from years of experience what contributed to the variation that could be observed in this kind of event. Then we captured reasons for why the longest duration was possible. We recorded these rationales for extreme outcomes of the event in a table like Table 11-1 (which is truncated here for the purpose of brevity). Notice, too, that the SME assigned a rank of importance to three of the extreme causes on either side, with a rank of 1 representing the most important factor in the SME's mind.

Table 11-1. *Example of Construction Engineer Acting as SME*

Rank	Reasons for Low	Reasons for High	Rank
1	Single source contractor.	Lack of market availability of steel.	1
2	Good weather permits favorable work conditions.	Inclement weather creates unfavorable work conditions.	2
3	Working with preferred supplier to use readily available steel shapes.	Multiple bidders.	3
	Use of assumed equipment loads for early design.	Coating used: galvanized and epoxy.	

In the case of discrete binary events, we follow a similar process, except that we document rationales for why the event might occur as defined or not. For example, a sales manager and her bid team were called on to provide a probability that their company would win a competitive bid response to a request for proposal from a government agency. At first the team expressed a high degree of confidence that the company would win the bid based on the strength of their market reputation and the high degree of quality of service they provided. Given this potential overconfidence, I asked them to think first about why the company might lose the bid. After we documented the rationales for the pessimistic outcome (losing the bid), we captured the rationales for the optimistic outcome (winning the bid). The results are given in Table 11-2.

Table 11-2. *Rationales for Outcomes*

Reasons for Bid Loss	Reasons for Bid Win
• We do not have good customer intimacy at multiple levels in the customer organization and have not directly vetted our solution with them.	• Our management and technical approach address the customer's needs and has been indirectly vetted (e.g., via Q&A, through consultants, industry briefs/interchanges, through team members) with the customer. Initial feedback seems positive.
• We have limited relationships with the customer and depend on open sources, public forums, and meetings to get information about their budget, underlying problems, tendencies, and needs.	• We have sufficient time to show the customer key elements of our solution and obtain direct feedback.
• We understand some of the parameters of the opportunity but knowledge gaps cause some uncertainty in our offer or competitive position.	• We have most of the facilities required to execute this program. What is left to fully execute, we can obtain later as it will not be needed in the early stages.
• We are missing an experienced and qualified program manager, technical leads, or appropriate SMEs, so we must rely on subcontractors to fill these positions.	• We will price this to win the business, even if it means a razor-thin margin, because it will provide an entry point for future business.
• A single award is expected, and there are multiple bidders. The odds are against us just by the number of bidders.	• We know the competitor's approach and have one that is significantly better with customer recognized discriminators.
• We have experience with similar programs; however, added effort is needed to clearly demonstrate relevance to receive good customer ratings on the bid.	
• We only have a rough idea of the customer's budget based on open source information.	
• This is new capability for a new customer. They might not have the full competency to assess the best offered solution.	

Measure the Range of Uncertain Events with Probabilities

Before proceeding with quantifying the uncertain event with the SME, assure the SME that the purpose of the interview is not to acquire a commitment or a firm projection about the outcome. You are merely trying to measure the overall uncertainty you face with the event. You should also assure the SME that you are not asking the SME to know the probabilities you are trying to assess as if probabilities are a thing or an intrinsic property of an event. Rather, you want to determine, by comparison to a defined lottery or game, the odds at which the SME is indifferent to choosing to play the lottery to win a prize or calling the outcome of the event in question to win a prize. The point of indifference for them is their subjective degree of belief—the probability—that the event will occur as defined.

There are generally two kinds of uncertain events that an SME will be called on to assess: discrete binary events and continuous events.

Discrete binary events are those that produce either a TRUE or FALSE state to describe the outcome of the event, that it either occurred or did not. Examples of this kind of event could be "win the deal," or "Phase 1 succeeds," or "Plaintiff claim is dismissed." The SME will assess the probability of one of the states. The probability of the complementary state must be 1 - Pr(state) by definition of the laws of probability. Under no circumstances can an SME assess the probability of one state, say TRUE, to be Pr(TRUE) = 0.25, and then assert that Pr(FALSE) = 0.55 or Pr(FALSE) = 0.95 without revealing a mental incoherence about the definition of the event. If this does occur, explore what is missing from the definition that might have led the SME to such incoherence.

Continuous events are assessed across an effectively infinitely divisible range with several probabilities, often at the 10th, 50th, and 90th percentile intervals. Examples of these kinds of events might be "capital cost of the reactor," or "duration of activity 25," or "sales in quarter 1." The p10 value is the value of the event at the 10th percentile for which the SME believes there is yet a 1/10 chance that the event outcome could still be lower. Conversely, the p90 value is the value of the event at the 90th percentile for which the SME believes there is yet a 1/10 chance that the event outcome could still be higher. The p50 value (or the median) is the value at which the SME considers that the outcome faces even odds or a 1/2 chance that the outcome could be either higher or lower.

Discrete Binary Uncertainties

Tell the SME he or she will have the hypothetical opportunity to win a prize by opting to play one of two games: Either the SME correctly calls the event occurring, or he or she can play a lottery by pulling a blue token from an urn or spinning an arrow on a wheel of fortune such that it lands on a blue sector for success. The choice to accept either the lottery or to call the outcome of the event is based on the odds presented for the lottery. If the SME thinks the odds of winning the lottery exceed the perceived odds of correctly calling the outcome of the uncertain event, then he or she should play the lottery. On the other hand, if the odds of winning the lottery are lower than the perceived odds of calling the event, then the SME should call the event. After a choice is made, the odds on the lottery are updated such that it should be more difficult to make a clear choice. This process is repeated until the SME is indifferent to choosing either game to win the prize.

The specific procedure goes as follows, using a method of halving distances to converge on the desired probability:

1. In the first iteration, produce a random number between 0 and 1, but never 0 or 1. Let Pr_1 be the initial probability of winning the lottery.

2. If the SME prefers the lottery, the lottery probability is too high. Update the next probability to half of the previous probability: $Pr\_2 = Pr\_1 / 2$.

3. If the SME prefers calling the event, the lottery probability is too low: $Pr\_2 = (1 + Pr\_1) / 2$.

4. If the SME starts with the lottery and continues to choose the lottery, then $Pr\_i+1 = Pr\_i / 2$. If the SME starts with calling the outcome and continues to call the outcome, then $Pr\_i+1 = (1 + Pr\_i) / 2$. However, as soon as the SME switches the game to pull back on overshoot in the last iteration, the probability updates as $Pr\_i+1 = (Pr\_i* + Pr\_i) / 2$, where Pr_i* is the last highest probability before the SME switches from playing the lottery to calling the outcome, or Pr_i* is the last lowest probability before the SME switches from calling the outcome to playing the lottery. In this way, the bracket of probabilities narrows and converges on a value where the SME should be indifferent to switching again.

5. Repeat Step 4 until the SME is indifferent to choosing either game to win a prize.

The following steps demonstrate the session with the previously described sales manager who assessed the discrete binary probability of winning a competitive bid for a government contract. Each step is illustrated by a result from this online tool[1] to facilitate the visualization of lottery probabilities on each iteration as the SME converges on an indifference point. Recall that the bid team initially expressed a high degree of confidence that their company would win the bid. After going through the rationale eliciting process, however, they realized that there might be more strong reasons for why they might lose the bid rather than win it; therefore, I switched the probability for them to assess to the probability of losing the bid to continue to counteract their initial overconfidence.

In the first iteration (Figure 11-3), a random value between 0 and 1 was selected, and the "win" state of the urn and the wheel of fortune are initialized with a proportion that matches the probability. So, the question now for the SME was this: Would you rather play the depicted lottery to win a prize, or would you rather call the outcome of the bid to win a prize (not win the bid)? Based on her internal sense of the weight of the rationales, the SME believed that the first iteration represented too low a value to play the lottery, so she chose to call the event, meaning that given her current information she believed there was a higher chance of losing the bid. Through the method described previously, the probability was updated to $Pr\_2 = 0.86$ (Figure 11-4). Again, the SME believed that the second iteration represented too low a value to chance the lottery, so she chose to call the event outcome. The probability was updated to $Pr\_3 = 0.93$ (Figure 11-5). On the fourth iteration, the SME believed the last update overshot the probability, so she chose to switch the game to play the lottery. The probability was then updated to $Pr\_4 = 0.89$ (Figure 11-6). At this point, the SME felt indifferent to choosing a game or calling the event outcome to win a prize, so she stayed with $Pr\_4 = 0.89$ as the probability of the event. Now, we knew the probability of winning the bid was $1 - Pr\_4 = 1 - 0.89 = 0.11$. Now the team had to carefully consider whether pursuing the bid was worth the effort at all; but if they did pursue it, they needed to plan how they might significantly improve their odds.

---

[1]http://www.incitedecisiontech.com/probcal.shtml

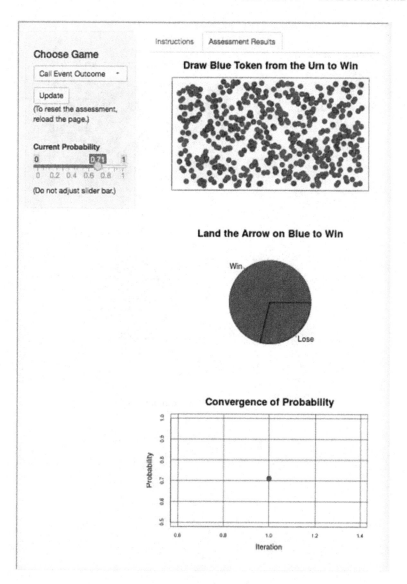

Figure 11-3. *The first iteration,* Pr_1 = 0.71, *was drawn randomly*

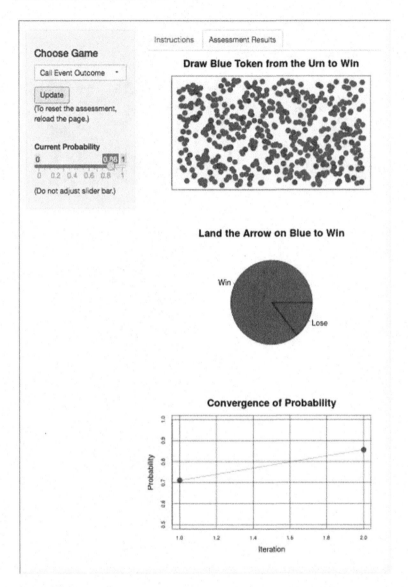

Figure 11-4. *The SME chose to call the event so the probability was updated to*
$Pr\_2 = (1 + 0.71) / 2 = 0.86$

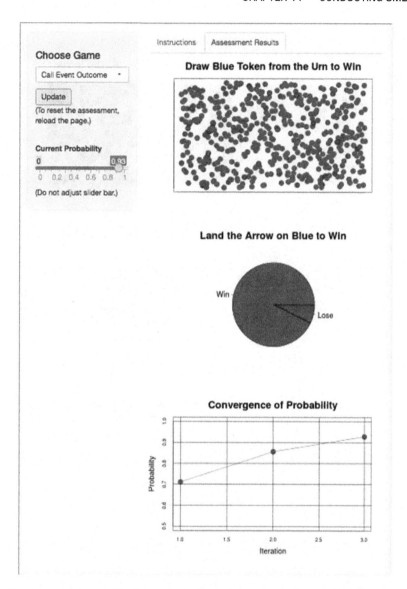

Figure 11-5. *Again the SME chose to call the event so the probability was updated to* `Pr_3 = (1 + 0.86) / 2 = 0.93`

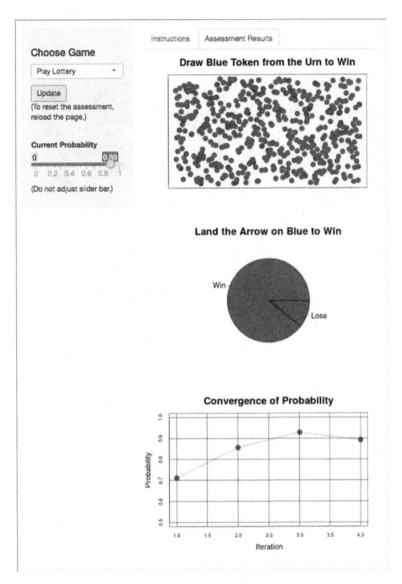

Figure 11-6. *On the fourth iteration, the SME chose to play the lottery, so the probability was then updated to* Pr_4 = (0.86 + 0.93) / 2 = 0.89. *At this point, the SME was indifferent either to playing the lottery or calling the outcome of the event, so the process was stopped.*

THE TROUBLE WITH SALES FORECASTS

Sales forecasts usually depend on the probability of winning a deal (i.e., Pr(win)). Think about the incentives most salespeople have, and ask yourself if the structure of incentives influences their ability to produce unbiased probabilities. Of course it does, but it does so according to the maturity of the salesperson. Younger salespeople tend to be overly optimistic. To correct for this optimism bias, I usually ask a salesperson to assess the probability of losing a deal. The probability of winning is Pr(win) = 1 - Pr(lose). More mature salespeople tend to sandbag their assessments of Pr(win). To correct for this, I ask the salesperson to describe how he or she usually produces a Pr(win) assessment and in what way he or she might adjust it before reporting it to the sales manager. In this case, I just assess the Pr(win). In my experience, however, most salespeople err on the side of optimism. Therefore, it's usually a good idea to start by asking how the Pr(win) is assessed. If sandbagging is not described, default to assessing Pr(lose).

Sometimes you will need to assess discrete events of more than two possible outcomes. These are the [A, B, C, ...] kinds of events. This is the general case of the binary discrete event, and it can be assessed in a similar manner as the binary one. Simply treat the A case as True, and assess its probability. The complementary probability is the residual probability that categories [B, C, ...] could occur. Now, assess the probability of B occurring (excluding A from the list). When the SME prefers to call the outcome, use this formula:

Pr_i+1 = (Pr_residual + Pr_i) / 2

instead of

Pr_i+1 = (1 + Pr_i) / 2.

Keep repeating this process until only one discrete category remains. The probability associated with this category will be 1 - (sum of all prior residual probabilities).

Continuous Uncertainties

As I have already alluded, the process of assessing the continuous uncertainty is very similar to that of assessing the discrete uncertainty, except that it usually includes three segments. However, instead of assessing a probability that an event will occur, the SME will be called on to provide an assessment for an 80th percentile prediction interval; that is, the range of values for which the SME posits an 80% degree of belief that will bracket the actual outcome.

Before beginning this process, avoid any inclination or suggestion to start with a "most likely" case. This just leads the SME to start with a biased anchor.

1. Start by assessing the P10 value. Tell the SME you are going to find the value for which he or she is indifferent to betting against a lottery with a 1/10 chance of winning or betting that the actual outcome will be less than his or her value. Ask the SME for a ballpark figure, order of magnitude starting point. Let this be P10_1.

2. If the SME prefers the lottery with a 1/10 chance of winning, that means that the P10_1 is too low.

3. Double this number: P10_2 = 2 * P10_1.

4. If the SME prefers betting that the actual outcome will be lower than P10_1 compared to taking the 1/10 chance lottery, that means that the P10_1 is too high. Take half of that number: P10_2 = P10_1 / 2.

5. Update the values on successive iterations such that the last high and last low values bracket the largest movement of the new value: P10_i+1 = (P10_i* + P10_i) / 2, where P10_i* is either the last high if the value needs to move up, or the last low if the probability needs to move down. In other words, you should never go back higher or lower than the last high or last low.

6. Repeat Step 4 until the SME is indifferent to choosing either game to win a prize. This value is the P10 associated with the 10th percentile probability.

7. Repeat Steps 1 through 5, but instead of betting on the actual value being less than the iterated value, the SME will bet on the actual value being greater than the iterated value. The indifference value is the P90 associated with the 90th percentile probability. The P90 should never be less than the P10.

8. The P50 value can be found in much the same way as the P10 value is found in Steps 1 through 5, except here you will use the P10 and P90 values as brackets, and the goal is to find the P50 value such that the SME is indifferent to playing a lottery with a 50% chance of winning (the odds of a coin toss) versus betting that the outcome will be higher or lower than some value between the P10 and P90. Be sure that the SME does not merely evenly divide the difference between the two endpoints.

9. To get an approximately full range (~99.9%), use the following formulas to find a virtual P0 and P100. These are the endpoints used in the BrownJohnson distribution described in Chapter 3.

    ```
    P0 = 2.5 * P10 - 1.5 * P50
    P100 = 2.5 * P90 - 1.5 * P50
    ```

If the SME wants to adjust the shape, he or she can do so either by moving the P10, P50, or P90, or by truncating the P0 to a position closer to the P10 or the P100 to a position closer to the P90. Typically, I do not recommend adjusting the P0 or P100 unless there is a systematic reason to impose a constraint, such as the actual value cannot go below zero or it is restricted from going above some number by policy, physical, or logical constraint. Otherwise, let the symmetry in the relative positions of the P10, P50, P90 determine the P0 and P100.

In the case presented earlier in which the construction engineer was called on to assess the duration of the structural steel construction for a chemical reactor, the engineer, reviewing the list of causative factors to the event duration, revealed his first indifference point for the 80th percentile prediction interval at P10 = 75 days. Then he revealed his P90 = 150 days. Finally, he was indifferent to betting whether the outcome would either be higher or lower than P50 = 107 days. Once we obtained these values, we used them to simulate distributions that allowed him to visually understand the implications of his assessments (Figure 11-7). Looking at the graph, though, made him realize that there might still be room for more upper tail. He then adjusted the P90 to 170 days. By taking this approach of exploring the edges of his belief about potential outcomes, we avoided an anchoring and adjustment bias. After the assessment ended, I asked the engineer what his original thoughts were about the duration distribution, and he conveyed to me that he actually had been concerned about the recent delay on a similar project. His original inclination was to suggest a "most likely" duration of 130 days with a ±20% variation (i.e., 104 days – 130 days – 156 days). The thinking process here helped him realize that not only might he have been too pessimistic, but that he was overly precise in his pessimism. The final assessment was much more accurate to the overall range of realistic durations, yet it still retained the possibility for his justified pessimism. You can use the online tool found at http://www.incitedecisiontech.com/distcal.shtml to help SMEs visualize the full range of their uncertainty assessments, too.

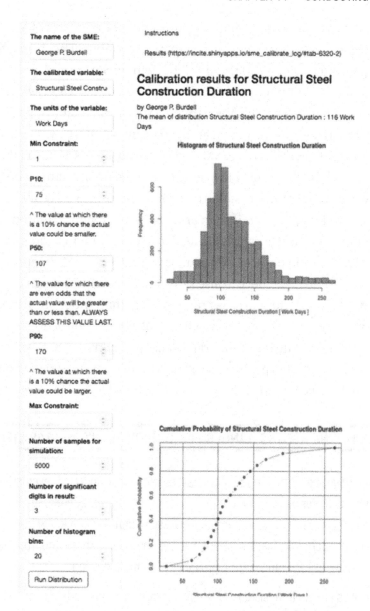

Figure 11-7. *The online continuous uncertainty calibration widget allows you to visualize the range of uncertainty implied by the the 80th percentile prediction interval supplied by an SME. Constraints can be supplied to override the natural P0 and P100 values, if necessary.*

It is absolutely imperative that you follow this sequence of events of assessing the outer values first and the interior value last, with no exceptions. The reason for this is that by assessing the outer values first, you will be helping SMEs think about events beyond their initial inclination to consider; thus, you will be avoiding availability bias. By assessing the inner value last, you will be helping SMEs avoid anchoring. Unfortunately, the all too common practice for producing ranged estimates starts with a "best guess" that is then padded with a ±X% that the SME believes represents a reasonable range of variation without thinking about how probable this range might actually be (it could be either grossly overstated or understated). The best guess is usually anchored in a biased direction, and the padding is just an arbitrary rule of thumb. Biased and arbitrary thinking are not mental characteristics that lead to accuracy in measurements.

Remember, although you are seeking accuracy with the best information you have at hand, you should not try to achieve a false sense of confidence through unwarranted precision by coaxing the SME for narrower ranges rather than wider ones. In fact, the wider the range you assess, the more likely you will take into account events outside the original biased inclination. Furthermore, you should avoid unnecessarily overworking the assessments to get the numbers perfect. The goal is to set bookends to the potential range of outcomes given the quality and limits of the SME's knowledge at the time of assessment.

WHY DON'T WE ASSESS RANGES FOR BINARY EVENT PROBABILITIES?

As an aside, you might be wondering why we assess ranges for uncertain events but only single point values for binary events. After all, you probably think one can't know the probability that an event will occur with greater precision than you can know the range of a continuous outcome. Certainly, we should put a range on the probability, too, you might think. The reasons we don't assess ranges for a probability are the following:

1. Probabilities are not intrinsic properties of natural events. Instead, they are subjective expressions about our degree of belief that an event will occur given all the information we can bring to bear on our consideration. If we put a range around the probability that an event will occur or not, we are revealing our internal incoherence about our degree of belief. Then, if we are willing to place a degree of belief around a degree of belief, should we not also be willing to place a degree of belief around each of our bracketing degrees of belief, and so on, and so on? Ultimately, we fall into an infinite regress that prevents us from finding a certain point of indifference between betting on two alternate lotteries of equivalent value.

2. When we assess the range of a continuous uncertainty, we can think of that range as a series of tiny little contiguous, nonoverlapping, mutually exclusive buckets of outcome. If any one of the little buckets occurs, that automatically excludes the other buckets from simultaneously occurring. The probabilities we assess for the three points (0.1, 0.5, 0.9) are actually cumulative probabilities that all the little buckets up to each point can occur. If we were to perform a reverse sum of the cumulative probabilities, the differences would be the probability that a particular bucket would occur or not. Any one little bucket we focus our attention on represents a binary event, which is the limiting case of assessing the probability of any binary event. In the continuous event, we are mapping the point of alignment of all our preassigned intangible degrees of belief on a range of unknown but tangible outcomes. In the singular binary event, we are assessing the unassigned intangible degree of belief that a specific preassigned tangible outcome will occur. In either case, a bucket gets only a single probability for us to maintain mental coherence about measurements we make on the world.

IT'S YOUR TURN

Try playing with the process outlined in this tutorial on specific events with friends and colleagues as the SME. For example:

- What will the price be of IBM stock (or any other equity you prefer) on a specific date?

- What year did Attila the Hun die? (Answer: 453 CE)

- What is the volume of Lake Erie? (Answer: 116 cubic miles; 480 km^3)

- What is the height of the tallest building in the world? (Answer: Burj Khalifa, Dubai, United Arab Emirates, 828 m, 2,717 ft., 163 floors)

- What is the airspeed velocity of an unladen swallow?

Before you attempt to assess any of these, think about specific clarifications and distinctions that need to be made to satisfy the clarity test. Keep track of how often the real value falls between your P10s and P90s.

Document the SME Interview

You will most likely find it helpful to document the content of the SME interview in a table like that shown in Figure 11-8.

Uncertainty Name:		Units:	
Definition:			
Name of SME: George P. Burdell		Date of Interview:	1-Jan-17

Reasons for Low Outcome:	Rank	Reasons for High Outcome:	Rank
1	1		1
2	2		2
3	3		3
4			
5			
6			
7			
8			
9			
10			

P10:	P50:	P90:

Figure 11-8. *SME interview table*

It's Just an Opinion, Right?

After reading this, you might conclude that all we're doing in this process is getting someone's opinion, that it's all still very subjective. You would be right. However, the process outlined here is objective, reproducible, and transparent.

1. *Objective*: There's no argument what the process requires, which effectively operates like a method of systematic variation, albeit as a virtual experiment in the mind of an expert.

2. *Reproducible*: Sufficiently experienced facilitators can assess different SMEs on the same uncertainty and frequently produce highly coherent assessments (although you can sometimes elicit widely divergent measurements, which can indicate that the subject matter is still under contention, or one of the SMEs is not nearly as well calibrated as the other).

3. *Transparent*: Rationales are clearly defined and documented for later inspection and auditing.

Even though we adhere to an objective process, the subjective nature of the information we elicit should not scare us away from using it. After all, even the best empirical data carry with them many subjective characteristics, namely, the methodological theory that was appealed to for systematically constructing the information, relaxations about the theoretical assumptions that the method specifies, and the subsequent choices that are made about which data are retained and which are discarded, to name a few. Of course, good empirical studies should find ways to eliminate systematic bias and document the ways in which they accomplish that goal or not. When we can use high-quality studies, we should. Unfortunately, though, empirical studies might not even be possible, as in the case of considering the opportunity costs between counterfactual or mutually exclusive hypothetical scenarios. The only way to get unbiased information in those cases (which represents most business case analyses) is to appeal to an omniscient, benevolent clairvoyant. So, in the absence of the best empirical data or omniscient reports from a benevolent clairvoyant, we must start somewhere. Expert judgment provides information that is reasonably constructed, and it is constructed in such a way that we can test the relative sensitivity of dependent figures of merit in our analysis to it. Ultimately, this points the way to the value of information analysis, which sets the research budget for updating our subjective SME prior with, possibly, improved empirical information.

CHAPTER 12

Kinds of Biases

The field of behavioral economics has catalogued more than 100 biases and heuristics that can obstruct clear thinking or produce cognitive illusions that can lead us to misperceive the world as it really is. Rather than produce an exhaustive list here, I provide a short list of the most important ones I seem to encounter most frequently.

- *Anchoring:* Using the first "best guess" as a starting point for subsequent estimating.

- *Availability:* Recalling values that are memorable, easily accessible, recent, or extreme.

- *Bandwagon bias or groupthink:* Conforming one's beliefs based on participation with a group norm. This is an especially strong bias if deviating from the group norm can result in ostracism or shaming, and participation in the group is considered a particularly valuable association. The experiments of Solomon Asch[1] in 1956 demonstrated just how powerful this effect can be.

- *Blind spot bias:* Failing to recognize that one's thinking is influenced by any biases at all.

- *Confirmation bias:* Selecting or using only information that supports a preexisting belief or perception.

- *Entitlement:* The SME provides an estimate that reinforces his or her sense of personal value.

---

[1]S. E. Asch, "Studies of Independence and Conformity: A Minority of One Against a Unanimous Majority. *Psychological Monographs*, 70.

© Robert D. Brown III 2018
R. D. Brown III, *Business Case Analysis with R*, https://doi.org/10.1007/978-1-4842-3495-2_12

- *Expert overconfidence:* Failure of creativity or professional hubris (e.g., "I know this information and can't be wrong because I'm the expert"). This often leads an SME to consider fewer possibilities associated with an event, leading back to false precision.

- *False precision:* Reporting anticipated outcomes with an unjustified level of certainty, usually as a single point estimate rather than a range.

- *Frame bias:* Failing to consider that there might be alternate ways to structure the context of a problem.

- *Incentives:* The SME experiences some benefit or cost in relationship to the outcome of the term being measured, adjusting his or her estimate in the direction of the preferred outcome.

- *Sand bagging:* Underreporting potential outcomes to appear heroic when better than anticipated outcomes materialize.

- *Selection bias:* Using information that is unintentionally filtered either by the method for collecting data or by an underlying mechanism (e.g., survivorship) such that the relevant population is not represented in an unbiased or randomized manner.

- *Unwarranted optimism:* Personal enthusiasm or a natural disposition to believe that desired outcomes will most likely occur; or, inflating initial estimates of desired outcomes to appear more effective than is warranted.

Familiarizing yourself with these biases will help you identify opportunities to gently challenge an SME's thinking in the elicitation process. For example, several years ago I helped a chemical manufacturing company develop a cost and schedule forecast for a new kind of chemical reactor. In an interview with an engineer about the possible duration required to construct one of the key mechanical structures, he began his discussion by telling me about a recent significant schedule overrun on another project of a similar nature. The situation had caused no small amount of internal controversy as late fees were incurred, fingers were pointed, and professional careers took abrupt redirections. As you might guess, I recognized his verbal rumination as an expression of the availability bias. By guiding the engineer to recall other projects that actually occurred on or within schedule, I was able to allow him to think more clearly about factors that might lead to a favorable duration. As a result, the development team was able to design acceleration plans for the project while reporting a historically consistent go live date to their client. In the end, everyone was delighted by the outcome.

In another situation, a sales team asked me to support their development of a revenue forecast for their government contracting company. To save some time, they provided a list of their target opportunities, some of which were current engagements that were up for contract renewal, each with an assigned probability of deal closure. The most important target was their key client, to which they assigned a 95% chance of closing their contract renewal. The sense of entitlement was, to put it mildly, palpable. Someone actually stated, "They'd be crazy not to re-up with us, but we discounted this 5% just so we don't look too arrogant to the CFO." Based on some background conversations that I had been privy to over the previous year, I believed this group was most assuredly not considering some important threatening information because they were blinded by their sense of entitlement. I asked them to turn the question around. Rather than thinking about what the probability of closure was, I asked them, instead, to think about what would lead to their losing the renewal. That conversation was much more sobering. The resulting revenue forecast, which included this "negative thinking" applied to all the other opportunities, indicated the need for bridge funding. Unfortunately, they needed it. Fortunately, they were prepared.

I want to clarify, though, that recognizing a given bias does not guarantee that favorable outcomes will occur or that disasters will be avoided. These two examples do not demonstrate that more accurate assessments cause the anticipated outcomes, as I've actually witnessed people think causality works in this manner. Rather, you will be better prepared to develop contingency plans for undesirable outcomes (or exploitation plans for the desirable ones) that will be uncovered by acknowledging the bias that would have otherwise kept these potential outcomes veiled from consideration. By recognizing thought-limiting biases, you can actually take more constructive steps to realize a future you or your clients desire.

A comprehensive list of cognitive and motivational biases can be found at the Behavior Economics Group web site (`https://www.behavioraleconomics.com/mini-encyclopedia-of-be/`). I also recommend taking in the delightfully entertaining and informative web site and podcasts of David McRaney at You Are Not So Smart: A Celebration of Self Delusion (`https://youarenotsosmart.com`). McRaney delivers engaging discussions about the biases and delusions that shape all our thinking.

PART 4

Information Espresso

Use value of information to make clear decisions efficiently

## CHAPTER 13

Setting a Budget for Making Decisions Clearly

Imagine standing beside your automobile at the starting line of a scavenger hunt and obstacle course that spans the continent from New York to Los Angeles. If you win, you will increase your income by $1 million. The trip does not come without its costs and risks, though. You will be responsible for the cost of food, fuel, and lodging. The route you take will bear a conditional effect on the travel costs as well as the cost of time. Every mile on the road will expose you to potential mechanical failures, the cost of repair and lost time, run-ins with the local gendarme, and possible roadway death. In your planning before the race (if you choose to plan at all), would you know how to choose the route that certainly results in the highest net gain?

The answer, of course, is "No," simply because many of the costs you will incur cannot be known until they are actually incurred. However, each cost you can anticipate will likely fall along a distribution of outcomes. By distribution, I don't just mean the range of the outcome, but also the likelihoods across the range.

Now, let's make the game more interesting. What if the winning purse is conditionally related to various outcomes during the race. Maybe you win more for crossing the finish line at earlier times. Maybe you gain or lose points related to the quality of the items you return that are specified in the scavenger hunt. Maybe each one of these opportunities to increase your earnings also affects the costs you might incur.

The net effect of all the uncertainties you face implies that the net value of any route you choose will likely overlap with the net value of the alternatives. Although one route might have a tendency that is better than the alternatives, the very notion that the potential range of the net values overlaps tells you that the choice is not necessarily

© Robert D. Brown III 2018
R. D. Brown III, *Business Case Analysis with R*, https://doi.org/10.1007/978-1-4842-3495-2_13

clearly unambiguous. There is some probability that even if you took the best average route, depending on how the slings and arrows of fortune fall, in a parallel universe you might experience a more desirable outcome by taking the route with the next best implied average value, or the next, next best.

What if you could improve your confidence in the route you choose by gaining better information about the potential outcomes of the uncertainties? How much would that be worth to you? What is the value you would place on the improved information? What is the most you would pay for it? How would you prioritize the open questions to which you need answers?

The preceding example serves as a metaphor for many complex business decisions, ones that include numerous uncertainties, risks, and competing goals and preferences. Often these decisions represent an unfolding series of questions that need to be answered as you move forward from start to finish. Returning to our scavenger hunt example, the first question you had to answer was whether the race was a good use of your time and talent to create value (which might be something other than just monetary income, such as having an exciting and memorable experience that confers a certain amount of bragging rights for having participated) compared to the other alternatives you had available to you. Once you committed to the decision to participate, then you had to decide which strategy to take in the execution of the race. Of course, few things ever go exactly as planned, so even during the course of the race you might pivot your current approach or abandon it altogether. Making decisions in a business context is not so different.

When we need to make a decision between major courses of action and need to answer 10, 20, or 100 questions that could distinguish the best way to go, how should we proceed?

- We could trust our gut and go with our intuition.

- We could gather more data and information.

- We could do some experiments.

Each one of those approaches imports some additional risk into what is already a risky gambit.

Trusting our gut usually exposes us to unanticipated outcomes that we could have anticipated beforehand. We are not going to anticipate every possible outcome, but there are many that we can. Being prepared in advance demonstrates a respect for the fiduciary responsibilities we have been entrusted with as directors of someone else's capital or concerns.

Gathering more information sounds and feels responsible. However, we often don't know which information to gather and how much to gather, parse, analyze, and interpret. The effort associated with these activities can easily be misappropriated on the wrong trails in search of the desired information. Even worse, the information gathering process sometimes leads into a never-ending foray that functions as a psychological excuse for delaying commitment to important decisions. There's always one more study to read, one more bit of reconnaissance to obtain, each accompanied with a nagging, almost neurotic, sense that we still don't know quite enough to make an informed decision.

Experiments often yield very important information; often they don't, for a host of reasons. One of the biggest risks with experimentation is very similar to that of seeking more information: Experiments can be expensive and can focus on phenomena that yield little insight into how to make a better decision. Obviously we cannot predict with certainty what the outcomes of our research will reveal; if we did, there would be no need for research. Developing business simulations (like that described in Chapters 2–4 of this book) can be expensive, too, although they are usually much less expensive than real-world experiments. This is usually because analysts become enamored with their models, pursuing functional complexity and detail, never quite getting to the point where actionable insights are revealed. The delay to action caused by the pursuit of irrelevant details is realized through the attrition of opportunity value.

Of the three preceding choices, I tend to believe that gathering information and conducting experiments are more fruitful than merely trusting one's emotional inclination. Given that, ideally we would like to know which pathways of information to explore or experiments to conduct that yield the most insightful results with the least expense of time and money. In other words, if we are going to seek more and better information to make important decisions with less ambiguity, before we do so, we'd like to know a prioritization of pathways to explore that gives us better answers with the most efficiency for the use of resources employed in the information gathering process. So, as I said, we cannot predict with certainty what the outcomes of our research will reveal, but in all likelihood there is a more-or-less rational prioritization of research efforts that lead to better returns, regardless of whether the least favored research thread ultimately turns out to be the most valuable. Value of information (VOI) provides a quantitative and cognitive instrument that helps us arrive at that prioritization. This tool is so powerful that it's like having a kind of clairvoyant at our disposal, and for that reason, VOI is often referred to as the value of clairvoyance.

VOI does not tell us which decision we should make. Rather, it tells us the maximum rational amount of the potential value associated with the best decision (as indicated by expected value before we have perfect information) we should be willing to give up to buy the information that helps us make an unambiguous decision. This maximum value is the budget we set for improving our current state of knowledge so that we can make a clear decision.

What You Will Learn

Part 4 of this book teaches decision makers how to tame decision ambiguity by establishing an efficient research budget used to clarify complex decisions under uncertainty.

The first section of this book set up a framework in the R programming language by which decision makers can analyze the uncertainty and risk associated with complex business opportunities. The middle sections set up a framework with three planning tools to create different strategic alternatives to capture the value of a complex business opportunity and provided a way to quantify uncertainty from SME knowledge. This section teaches you how, again with the support of the R language, to compare the effects of uncertainty across those strategic alternatives in a time- and resource-efficient manner.

In this section you will learn the following:

- The meaning of VOI.

- How to calculate the VOI in R.

For Your Consideration   How often have you told yourself, "I could have made a better decision if I had more information"? But, how often have you asked yourself if you paid **too much money**—or spent **too much time** looking—for additional information to clarify an impending decision?

CHAPTER 14

A More Refined Explanation of VOI

Now that we have a notional bird's-eye view of VOI, let's explore how to quantitatively evaluate it from a simple instructive example. We start with a decision tree example, then move to the now familiar influence diagram representation.

The Decision Tree

Imagine that you are a member of a strategic planning group in a biomedical device company. After learning how to generate creative decision strategies, your group is considering two strategic pathways to take a new device to market. You have determined that the value of each strategic pathway is dependent on only one currently unrealized uncertainty—let's call this a critical uncertainty—with a distribution that is conditional on a given pathway. Maybe the uncertainty is the addressable population served by the device or its range of efficacy (either determined by the engineering design chosen for a given pathway), or maybe it's the capital expenditure. Whatever the case, depending on which pathway you decide to take, the critical uncertainty will realize across a range of potential values and associated likelihoods. For any value that manifests, the chosen pathway will result in a net commercial value (i.e., NPV) to the company.

For the sake of discussion, let's assume the following average NPVs would be evaluated on the cumulative probability p10, p50, (median), and p90 outcomes of the conditional uncertainty:

- NPV(p10, p50, p90 | path A) = $ (140, 170, 230) M

- NPV(p10, p50, p90 | path B) = $ (120, 150, 210) M

© Robert D. Brown III 2018
R. D. Brown III, *Business Case Analysis with R*, https://doi.org/10.1007/978-1-4842-3495-2_14

To be clear, the numerical values are the average NPVs when the conditional uncertainty resolves to its (p10, p50, p90) values. These are not the (p10, p50, p90) values of the potential distribution of the NPV.

A decision tree provides a useful tool for clarifying the decision problem before us in terms of the sequence of events, relevant probabilities of uncertainties, and final outcomes. Here we see that your decision (starting with the green square) is whether to take one pathway (A) or the other (B). Then you face the potential outcomes of the uncertainty (the branches from the light blue circles) of concern conditioned by our choice of pathway. On the end of each pathway (red triangles on the end of each decision and uncertainty outcome), you realize the conditional expected (or average) estimate of NPV.

Notice in Figure 14-1 that each branch after an uncertain node has probability weights assigned to the specific branches associated with the quantile level of the uncertainty that could be realized. These are the extended Swanson-Megill probabilities frequently assigned to (p10, p50, p90) outcomes for analysis.

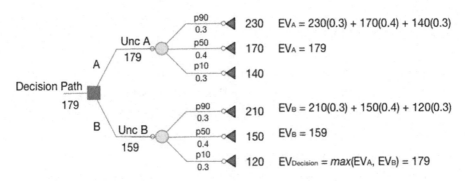

Figure 14-1. The initial decision tree that indicates the value of making a decision now based on our current information before we know the outcome of the relevant conditional uncertainty

Extended Swanson-Megill Probabilities   To quickly review part of our discussion in Chapter 3, the probabilities 0.3, 0.4, and 0.3 come from the extended Swanson-Megill[1] discrete weights that can be applied to the p10, p50, and p90 percentile values of a continuous distribution as discrete approximations of the distribution. The weightings are so derived to preserve the mean and variance of the original distribution, especially for those distributions with skew. Be aware that representing an uncertainty at different quantile values pX, pY, pZ will require different weights. The extended Swanson-Megill probabilities are applied to (p10, p50, p90) values only.

The decision tree "rolls back" the values and probability weights to produce the decision path expected value. We take the maximum of these expected values on a decision node to indicate one's selection of the prescribed pathway—assuming that decision makers want to maximize their value.

The following describes how the rollback process works. First, working from the right to the left, multiply the probabilities of each uncertain branch by the outcomes of each branch. Then add the products.

```
# Expected Value for Pathway A
EV_A = (0.3 * $140M) + (0.4 * $170M) + (0.3 * $230M)
EV_A = $179M

Expected Value for Pathway B
EV_B = (0.3 * $120M) + (0.4 * $150M) + (0.3 * $210M)
EV_B = $159M
```

Find the maximum value of the decision branches.

```
# Decision Value
EV_Decision = max($179M, $159M) = $179M
```

---

[1]MEGILL, ROBERT E., *An Introductionto Risk Analysis*, Petroleum Publishing Company, Tulsa, 1977.

Based on this approach, we can see that the prescribed pathway based on expected NPV alone would be to take strategic pathway A. That seems simple enough: Take the pathway that offers the best average value looking forward. After all, if you had better information you would use it to refine the decision path value forecast. However, you still can't help but feel a lingering sense of unease given the range of overlap in the pathway values. Regardless of what you might have been taught in business school, the best decision still seems a wee bit ambiguous.

Wait, though. What if you could get better information to "buy down" the range of uncertainty in the uncertainty of concern so that the decision was just clear enough to be unambiguous? How much would that information be worth to you? How much should you be willing to pay to get that information?

The question now for you is whether you will pay for certain information that will tell you if ongoing development of the investment project is a rational path. Now we modify the decision tree in Figure 14-1 to show that you know the outcome of the uncertainty for each decision before you continue down a decision pathway (Figure 14-2) and compare that to the case where you don't know (the previous decision tree). In other words, you compare the value of the decision opportunity with prior information to that of your current information. The difference between these two is the rational value you should be willing to pay for the information that makes the decision just unambiguous.

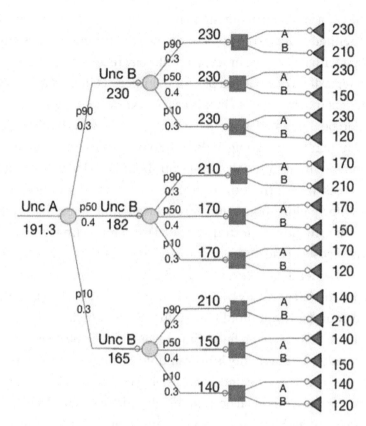

Figure 14-2. *The decision tree of Figure 14-1 reordered to reflect that the outcome of the relevant conditional uncertainty is known before making a decision*

By rolling back the decision tree again similar to the way we did before (I'll leave it to the reader to work out these values on their own), we see that the value of knowing the outcome of the uncertainty before you choose a committed pathway is equal to $191.3 million; therefore, the incremental value over not having prior information is $191.3 million - $179 million = $12.3 million. This is the VOI: the rational amount to pay for perfect information. That's right: perfect information.

Of course, the information you obtain might not be perfect. Maybe this information is obtained via a refined market study or a designed experiment. Because there is no such thing as a test with perfect accuracy, you will only receive imperfect information in the real world regardless of what you pay for it. Imperfect information is always less valuable than perfect information, so in this case we know that the VOI we originally determined should be thought of as the maximum amount you should rationally be willing to pay. Although I'm not going to delve into the more complicated aspects of imperfect information in this tutorial, I think you should be able to see that in cases where you incur costs to perform some experiment or test that produces less than perfectly precise information about critically uncertain outcomes, this is equivalent to buying imperfect information. Unfortunately, R&D and marketing groups frequently incur these kinds of costs in the forms of designed experiments or market research without knowing whether the cost exceeds a rational amount or not. Of course, no one requires you to pay the full perfect VOI, but if you know the VOI on a given uncertainty is greater than zero, rationally, you should be willing to pay something for the information.

Be aware that the VOI can often be $0 (but never less than $0). This is particularly true when the underlying uncertainty is not conditional on the decision that is made, which is usually the case when the uncertainty being considered is a commodity price, such as the market price of crude oil or a chemical feedstock. If the VOI is indeed $0, this is equivalent to saying that the value given prior information is the same as one's current state of knowledge. In this case, you should not spend any more time or money to resolve the outcome of the uncertainty and its contingent effects before you are exposed to it.

Some Preliminary R Code

Of course, working out the calculations for the simple problem just presented is a fairly trivial exercise, as one can work them out with paper and pencil. However, we need to understand how to set up the calculations for a more automated process in larger, more complex problems.

Calculating the decision pathway value with current information is quite simple.

```
# Mean NPVs corresponding to the p10, p50, p90 quantile values of the
# conditional uncertainty.
dec.val.A <- c(140, 170, 230)
dec.val.B <- c(120, 150, 210)
```

```
# The expected NPV of each pathway is the probability weighted average NPV.
# Using the extended Swanson-Megill weights for p10, p50, p90 quantiles.
prob.wts <- c(0.3, 0.4, 0.3)
dec.val.A.m <- sum(dec.val.A * prob.wts)
dec.val.B.m <- sum(dec.val.B * prob.wts)
```

```
> dec.val.A.m
[1] 179
```

```
> dec.val.B.m
[1] 159
```

Note that the number of pathways in the decision tree representing the case of having prior information is just the number of all the combinations of the possible branches given the way we have partitioned the outcome of each uncertainty into three quantiles. In this case, there are (number of branches)$^{\text{(number of uncertainties)}}$ = 3^2 = 9 combinations. The conditional value of each branch would simply be the Cartesian product of the weights assigned to each quantile.

```
# The Cartesian product of the weights.
probs.branch.matrix <- as.array(prob.wts %*% t(prob.wts))
```

```
> probs.branch.matrix
     [,1] [,2] [,3]
[1,] 0.09 0.12 0.09
[2,] 0.12 0.16 0.12
[3,] 0.09 0.12 0.09
```

The prescribed decision value on each branch of the decision tree can be found as the parallel maximum of each Cartesian coordinate of the decision values with current information.

```
dec.val.A.matrix <- array(rep(dec.val.A, 3), dim = c(3, 3))
dec.val.B.matrix <- t(array(rep(dec.val.B, 3), dim = c(3, 3)))
```

```
> dec.val.A.matrix
      [,1] [,2] [,3]
[1,]  140  140  140
[2,]  170  170  170
[3,]  230  230  230
```

```
> dec.val.B.matrix
     [,1] [,2] [,3]
[1,]  120  150  210
[2,]  120  150  210
[3,]  120  150  210
```

```
# Find the parallel max value of the two prior matrices.
path.max <- pmax(dec.val.A.matrix, dec.val.B.matrix)
```

```
> path.max
     [,1] [,2] [,3]
[1,]  140  150  210
[2,]  170  170  210
[3,]  230  230  230
```

The decision value with prior information is just the array sum of the product of the probs.branch.matrix with the path.max.

```
dec.val.prior.info <- sum(path.max * probs.branch.matrix)
```

```
> dec.val.prior.info
[1] 191.3
```

Finally, the VOI is the difference between the decision value with prior information and the decision that maximizes value with current information.

```
voi <- dec.val.prior.info - max(dec.val.A.m, dec.val.B.m)
```

```
> voi
[1] 12.3
```

That's the punchline. Calculating the VOI of a given conditional uncertainty for a binary decision reduces down to the following steps:

1. Isolate the critical uncertainty.

2. Determine the contingent effect on the value measure by holding the uncertainty fixed at some key branches (namely, some standard quantiles, like p10, p50, p90).

3. Find the probability weighted value of the prescribed max values for all the combinations of the conditional uncertainties.

4. Find the difference between Step 3 and the maximum mean value prescribed by current information.

One of the key preliminary tasks for calculating the VOI, though, is isolating the appropriate uncertainties for evaluation. Requisite business case analysis models can often include a half-dozen to three or four dozen uncertainties. Not only would it be tedious and error prone to write code that calculated the VOI of all the conditional uncertainties in a business case analysis, but it would also be unnecessary. This is because it's often the case that about three or four of those uncertainties pose a critical potential to lead to regret for choosing a decision based on maximum expected value only. Tornado analysis, which we first saw in Chpater 2, shows us how to isolate candidate critical uncertainties for VOI evaluation. We will cover that in the next chapter. Before we get there, let's review the value of using influence diagrams to replace decision trees as a visual guide to manage our state of knowledge of the effect of uncertainties and decisions on the measures of important values.

The Influence Diagram

Decision trees are useful tools for analyzing simple decision problems under uncertainty. However, their use with real issues poses a problem for analysts. As decision problems become more complex, decision trees often become decision bushes. Although software is available that can manage the quantitative effort, the cognitive effort required by analysts and the consumers of their analysis can become burdened by the morass of branches.

To get around the problem of visual complexity, we can use an influence diagram. An influence diagram visually displays the flow of influence (cause, relevance, or correlation represented by arrowed arcs) from certain types of entities (represented by nodes of different shapes) that are contextually related within a decision problem to some value we use to distinguish one decision path over another. Usually, the value measurement represents monetary value (i.e., NPV of a cash flow), but frequently other measures are used, such as time durations to important events, lives at risk, and so on. The value measure represents the thing you wish to minimize or maximize by making a choice or several coordinated choices.

INFLUENCE DIAGRAM NOTATION

In an influence diagram, arcs and nodes convey information about key aspects of the decision problem it represents. Node entities are distinguished by their shape (and sometimes color):

- *Rectangles (green)*: Decisions

- *Ellipses (cyan)*: Uncertainties

- *Double lined ovals (yellow)*: Intermediate calculations

- *Trapezoids (purple)*: Facts or constant values

- *Diamonds or hexagons (red)*: The objective or key value measure

- *Arrowed arc*: The flow of influence (relevance or conditional relationship) an entity bears on another

Think of the influence diagram as a space of analytic information you need to know to resolve the question: Which decision path should I prefer to maximize the likelihood of achieving the outcome I want? It forces us to account for a measurement of the outcome we want, explicitly known facts, the key decisions we need to make, and what we don't know with certainty as a condition of those decisions. In my mind, though, the greatest benefit of the influence diagram is that it functions as an accounting of the effects of uncertainty. Tornado charts, a type of two-way sensitivity analysis that emerges from the influence diagram logic, give us a clue about how to prioritize our analysis space in the most economically efficient way given the quality of the information we currently possess. The VOI is the quantitative prioritization of that analysis space with a maximum associated budget for which we should be willing to expend to remove our doubts that we are pursuing the best strategy among those we've conceived. Further, influence diagrams can eventually be converted into Bayesian networks, a more advanced tool in the business analytics toolbox.

The influence diagram for the original problem stated earlier is depicted in Figure 14-3.

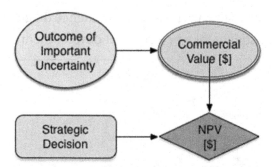

Figure 14-3. *The simple influence diagram for making a decision prior to knowing the effect of a relevant uncertainty on the value function*

If we include knowledge about the outcome before we make decisions, the diagram now looks like Figure 14-4.

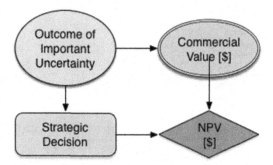

Figure 14-4. *The simple influence diagram updated to indicate that a decision is made after knowing the outcome of a relevant uncertainty*

The influence diagram by itself doesn't tell us explicitly how to perform any calculations, but as I laid out in the first part of this book, developing an influence diagram is a very important part of thinking through decision problems because it does the following:

1. Lays out the context of the decision problem at hand.

2. Clarifies the kinds of information needed to think through the problem.

3. Provides an abstraction of the problem while avoiding the visual complexities of decision trees.

4. Describes the general flow of calculations that will be needed from assumptions and constraints through intermediate calculations to the key outcome that we desire to measure.

For the remaining discussion, we refer to the influence diagram in Figure 14-5 that represents the value sought for a new product development through two different strategic pathways.

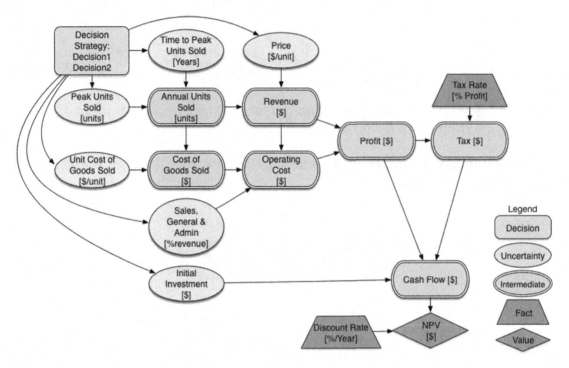

Figure 14-5. *The influence diagram of the simplified business case model*

Looking at this influence diagram you might wonder why the decision node connects to the various uncertainties. It might seem to imply that the decision causes their outcome. If that were the case, though, the uncertainties wouldn't be uncertain. What the diagram actually conveys is that the uncertainties are conditionally related to which decision choice is exercised. Think of it like this: Suppose you are making a decision about where to go on summer vacation, and you want to go someplace warm where you will likely experience the least rainfall and the most sun. If you go to the U.S. Southwest, you will probably enjoy a lot of sun and very little, if any, rain; however, if you go to Florida, you will mostly likely experience warm, sunny, muggy days. Unfortunately, you will also face various levels of daily rain showers that occur in the late afternoon. Your decision does not influence the variety of the weather outcomes, but the variety you experience will be conditionally related to the decision you make.

CHAPTER 15

Building the Simulation in R

As for the solution approach, you want to calculate the VOI for the problem displayed in our last influence diagram. To do so, we take the following algorithm approach.

1. Run the simulation for both decision strategies (or more as indicated) and calculate the mean NPV and NPV cumulative probability distributions. If one of the cumulative probability distributions does not exhibit strict dominance, then we know that sensitivity analysis might be helpful for identifying critical uncertainties.

2. Test the sensitivity of the mean NPV of each decision to the underlying uncertainties using tornado analysis and identify which uncertainties might cause us to regret taking the best strategy (on the basis of expected value) in contrast to the next best strategy.

3. Find the VOI as the difference between the improved NPV by exercising choice based on prior information and the best mean NPV from Step 1.

© Robert D. Brown III 2018
R. D. Brown III, *Business Case Analysis with R*, https://doi.org/10.1007/978-1-4842-3495-2_15

The Model Algorithms

Before we proceed, we need to set up a few important files that we will import into our R code. The first is a file that contains important functions, one of which is the business model that we will use in the simulation of the values expressed in the influence diagram. We use the following functions:

1. *CalcBrownJohnson*: Takes p10, p50, p90 assessments from an SME and generates a simulation of samples that spans beyond the p10 and p90 to include the virtual tails.

2. *CalcScurv*: Models the saturation of uptake into a population using a sigmoid curve. For example, this can be used to model how long it takes to reach maximum saturation into a marketplace.

3. *CalcBizModel*: The function that represents the business model reflected by the influence diagram.

4. *CalcModelSensitivity*: Takes a list of uncertainty simulation values and calculates the sequential sensitivity of the NPV returned by the CalcBizModel to specified quantiles (e.g., the p10, p50, p90) for each uncertainty.

Name this file Functions.R, and populate it with the following code.

```
CalcBrownJohnson <- function(minlim=-Inf, p10, p50, p90, maxlim=Inf,
samples) {
# This function simulates a distribution from three expert estimates
# for the 80th percentile probability interval of a predicted outcome.
# The user specifies the three parameters and the number of samples.
# The user can also enter optional minimum and maximum limits that
# represent constraints imposed by the system being modeled. These are
# set to -inf and inf, respectively, by default. The process of
# simulation is simple Monte Carlo with 100 samples by default.

    # Create a uniform variate sample space in the interval (0,1).
    U <- runif(samples, 0, 1)
    lenU <- length(U)
    # Create an index in the interval (1,samples) with samples members.
    Uindex <- 1:lenU
```

```r
# Calculates the virtual tails of the distribution given the p10,
p50, p90
# inputs.
p0 <- pmax(minlim, p50 - 2.5 * (p50 - p10))
p100 <- pmin(maxlim, p50 + 2.5 * (p90 - p50))

# This next section finds the linear coefficients of the system of linear
# equations that describe the linear spline, using... [C](A) = (X)
(A) = [C]^-1
# * (X) In this case, the elements of (C) are found using the values
(0, 0.1,
# 0.5, 0.9, 1) at the endpoints of each spline segment. The elements
of (X)
# correspond to the values of (p0, p10, p10, p50, p50, p90, p90,
p100). Solving
# for this system of linear equations gives linear coefficients that
transform
# values in U to intermediate values in X. Because there are four
segments in
# the linear spline, and each segment contains two unknowns, a total
of eight
# equations are required to solve the system.

# The spline knot values in the X domain.
knot_vector <- c(p0, p10, p10, p50, p50, p90, p90, p100)

# The solutions to the eight equations at the knot points required to
describe
# the linear system.
coeff_vals <- c(0, 1, 0, 0, 0, 0, 0, 0,
                0.1, 1, 0, 0, 0, 0, 0, 0,
                0, 0, 0.1, 1, 0, 0, 0, 0,
                0, 0, 0.5, 1, 0, 0, 0, 0,
                0, 0, 0, 0, 0.5, 1, 0, 0,
                0, 0, 0, 0, 0.9, 1, 0, 0,
                0, 0, 0, 0, 0, 0, 0.9, 1,
                0, 0, 0, 0, 0, 0, 1, 1)
```

```
# The coefficient matrix created from the prior vector. It looks like the
# following matrix:
          # [,1] [,2] [,3] [,4] [,5] [,6] [,7] [,8]
   # [1,]  0.0   1  0.0   0  0.0   0  0.0   0
   # [2,]  0.1   1  0.0   0  0.0   0  0.0   0
   # [3,]  0.0   0  0.1   1  0.0   0  0.0   0
   # [4,]  0.0   0  0.5   1  0.0   0  0.0   0
   # [5,]  0.0   0  0.0   0  0.5   1  0.0   0
   # [6,]  0.0   0  0.0   0  0.9   1  0.0   0
   # [7,]  0.0   0  0.0   0  0.0   0  0.9   1
   # [8,]  0.0   0  0.0   0  0.0   0  1.0   1

coeff_matrix <- t(matrix(coeff_vals, nrow=8, ncol=8))


#The inverse of the coefficient matrix.
inv_coeff_matrix <- solve(coeff_matrix)

#The solution vector of the linear coefficients.
sol_vect <- inv_coeff_matrix %*% knot_vector

X = (U <= 0.1) * (sol_vect[1, 1] * U + sol_vect[2, 1]) +
    (U > 0.1 & U <= 0.5) * (sol_vect[3, 1] * U + sol_vect[4, 1]) +
    (U > 0.5 & U <= 0.9) * (sol_vect[5, 1] * U + sol_vect[6, 1]) +
    (U > 0.9 & U <= 1) * (sol_vect[7, 1] * U + sol_vect[8, 1])

   return(X)
}

CalcScurv <- function(y0, t, tp, k = 0) {
  # This function models the sigmoid 1 / (1 + exp(-g*t)) with four analytic
  # parameters.
  # y0 = saturation in first period.
  # t = the index along which the s-curve responds, usually thought of as time.
  # k = the offset in t for when the sigmoid begins. Subtract k for a right
  #    shift. Add k for a left shift.
  # tp = the t at which s-curve achieves 1-y0.
```

```r
  this.s.curve = 1 / (1 + (y0 / (1 - y0)) ^ (2 * (t + k) / tp - 1))
  return(this.s.curve)
}

CalcBizModel <- function(Time, N, pus, ttp, p, sga, cogs, tr, i, dr) {
  # The function that represents the business model reflected by the
  influence
  # diagram.

        # Time = the time index
        # N = the number of simulation samples
        # pus = peak units sold
        # ttp = time to peak units sold
        # p = price, $/unit
        # sga = sales, general, and admin, % revenue
        # cogs = cost of goods sold, $/unit
        # tr = tax rate, %
        # i = initial investment, $
        # dr = discount rate, %/year

        init <- t(array(0, dim=c(length(Time), N)))
        ann.units.sold <- init
        revenue <- init
        profit <- init
        tax <- init
        cash.flow <- init
        npv <- rep(0, N)
        samp.index <- 1:N
        for (s in samp.index) {
          # annual units sold
          ann.units.sold[s, ] <-
            (Time > min(Time)) * pus[s] * CalcScurv(0.02, Time, ttp[s], -1)

        # annual period revenue
        revenue[s, ] <- p[s] * ann.units.sold[s, ]

        # annual period profit
        profit[s, ] <-
```

```r
    revenue[s, ] - (sga[s] * revenue[s, ]) - (cogs[s] * ann.units.
    sold[s, ])

  # annual period tax
  tax[s, ] <- tr * profit[s, ]

  # annual period cash flow
  cash.flow[s, ] <-
    (Time > min(Time)) * (profit[s, ] - tax[s, ]) - (Time ==
    min(Time)) * i[s]

  # net present value of the annual period cash flow
  npv[s] <- sum(cash.flow[s, ] / (1 + dr) ^ Time)
  }

# Collect all intermediate calculations in a list to be used for
other
# calculations or reporting.
calc.vals <- list(ann.units.sold = ann.units.sold,
                    revenue = revenue,
                    profit = profit,
                    tax = tax,
                    cash.flow = cash.flow,
                    npv = npv
                  )
  return(calc.vals)
}

CalcModelSensitivity <- function(unc.list, sens.q) {
  # unc.list = A list that contains the uncertain variables' samples used in
  #            the business model.
  # sens.q = a vector that contains sensitivity test quantiles

  # Create an index from 1 to the number of uncertainties used in the business
  # model.
  unc.index <- 1:length(unc.list)

  # Assign the values of the uncertainties list to a temporary list
  uncs.temp <- unc.list
```

```r
# Initialize a table to contain mean NPVs of the business model as each
# uncertainty is set to the sensitivity quantile values.
sens.table <-
  array(0, dim = c(length(unc.list), length(sens.q)))
row.names(sens.table) <- names(unc.list)
colnames(sens.table) <- sens.q

for (u in unc.index) {
  for (s in 1:length(sens.q)) {

    # Iterate across the uncertainties and elements of the sensitivity
    # quantile values and temporarily replace each uncertainty's samples with
    # the uncertainty's value at each quantile value.
    uncs.temp[[u]] <- rep(quantile(unc.list[[u]], sens.q[s]), samps)

    # Populate the sensitivity table with the mean NPV values calculated in
    # the business model using the values in the temporary uncertainty list.
    sens.table[u, s] <- mean(
      CalcBizModel(
        time,
        samps,
        uncs.temp$peak.units.sold,
        uncs.temp$time.to.peak,
        uncs.temp$price,
        uncs.temp$sga,
        uncs.temp$cogs,
        tax.rate,
        uncs.temp$investment,
        disc.rate
      )$npv
    )
  }
  # Reset the temporary uncertainty list back to the original uncertainty
  # list.
  uncs.temp <- unc.list
}
return(sens.table)
}
```

Note that the business model from the influence diagram is treated as the function `CalcBizModel()`. The reason I chose to do it this way is that I want to make sure that the logic is consistent among the decision pathways I consider, and I only want to pass relevant information (facts and uncertainties) to the model for the purpose of calculating VOI. If you want access to the intermediate results for other analytic uses and later financial analysis and planning, these values are returned as a list when the function runs. Be aware that because the model is a probabilistic simulation, the values in the returned list are simulation samples; use `set.seed()` if you want to ensure reproducibility in your results.

You might also notice that while reviewing the functions, and elsewhere in the tutorial, that I depend on `for()` loops for iteration blocks rather than the often prescribed `apply()` functions. Yes, the code runs more slowly than otherwise, but in the end I decided that goal of explaining the purpose of the algorithms was clearer using `for()` rather than `apply()`. If you are new to R and don't know what the `apply()` family accomplishes as a replacement for `for()` loops, you should learn that quickly. The overall speed of your code will drastically improve.

Next, we need a file that contains our various assumptions and uncertainty assessments. You can name this file `Assumptions.R`.

```r
# time horizon of the model
time.horizon <- 10 # yr
time <- 0:time.horizon
set.seed(98)

# number of samples used for the simulation.
samps <- 30000

tax.rate <- 0.35 # %
disc.rate <- 0.10 # %/yr

# Uncertainties have the following units.
# peak.units.sold, max units/year after year 0
# time.to.peak, years
# price, $/unit
# cogs, $/unit
# sga, % revenue
# investment, $ in year 0
```

```r
# uncertainty parameters for decision 1
dec1.uncs <- list(
  peak.units.sold = round(CalcBrownJohnson(0, 10000, 12500, 18000, , samps)),
  time.to.peak = CalcBrownJohnson(0, 2, 3, 5, , samps),
  price = CalcBrownJohnson(0, 90, 95, 98, , samps),
  cogs = CalcBrownJohnson(0, 15, 17, 20, , samps),
  sga = CalcBrownJohnson(0, 0.1, 0.12, 0.14, 1, samps),
  investment = CalcBrownJohnson(0, 100000, 115000, 150000, , samps)
)

# uncertainty parameters for decision 2
dec2.uncs = list(
  peak.units.sold = round(CalcBrownJohnson(0, 9000, 17000, 25000, , samps)),
  time.to.peak = CalcBrownJohnson(0, 2, 4, 6, , samps),
  price = CalcBrownJohnson(0, 95, 100, 103, , samps),
  cogs = CalcBrownJohnson(0, 16, 18, 21, , samps),
  sga = CalcBrownJohnson(0, 0.1, 0.13, 0.16, 1, samps),
  investment = CalcBrownJohnson(0, 135000, 155000, 200000, , samps)
)
```

Normally, I recommend keeping the parameter data for the uncertainties in CSV files, importing them through the read.csv() function as demonstrated in Chapter 2, and then assigning the uncertainty parameters to each uncertainty with the CalcBrownJohnson() function. In this case, however, the list of uncertainties is short for the purpose of demonstration, so to make things a little simpler and to focus more on the VOI calculation, I took a shortcut.

Now that we have the functional and data pieces in place to initialize and run our model, we can use the following code in, say, Business_Decision_Model.R, to accomplish Step 1 in our solution approach.

```r
# Import function and data source files.
source("/Applications/R/RProjects/Value of Information/Functions.R")
source("/Applications/R/RProjects/Value of Information/Assumptions.R")

# Run the business decision models and return sample outputs of NPV.
bdm1 <-
  CalcBizModel(
    time,
```

```
      samps,
      dec1.uncs$peak.units.sold,
      dec1.uncs$time.to.peak,
      dec1.uncs$price,
      dec1.uncs$sga,
      dec1.uncs$cogs,
      tax.rate,
      dec1.uncs$investment,
      disc.rate
    )$npv / 1e6

bdm2 <-
    CalcBizModel(
      time,
      samps,
      dec2.uncs$peak.units.sold,
      dec2.uncs$time.to.peak,
      dec2.uncs$price,
      dec2.uncs$sga,
      dec2.uncs$cogs,
      tax.rate,
      dec2.uncs$investment,
      disc.rate
    )$npv / 1e6

# Calculate the mean NPV of each decision and their difference.
m.bdm1 <- mean(bdm1)
m.bdm2 <- mean(bdm2)
diff.bdm <- mean(bdm2 - bdm1)

# Tabulate the average model results.
m.bdm <- array(c(m.bdm1, m.bdm2), dim=c(2, 1))
colnames(m.bdm) <- "Mean NPV of decision [$M]"
rownames(m.bdm) <- c("Decision1", "Decision2")
print(signif(m.bdm,3))
print(paste("The difference in mean value between strategies:", "$",
            signif(diff.bdm, 3), "M"))
```

```
# Plot the cumulative probability distribution of the NPVs.
par(mar = c(12, 5, 5, 2) + .05, xpd = TRUE)
legend.idents <- c("Decision 1", "Decision 2")
plot.ecdf(
  bdm1,
  do.points = TRUE,
  col = "red",
  main = "NPV of Business Decision",
  xlab = "NPV [$M]",
  ylab = "Cumulative Probability",
  xlim = c(min(bdm1, bdm2), max(bdm1, bdm2)),
  tck = 1
)
plot.ecdf(
  bdm2,
  do.points = TRUE,
  add = TRUE,
  col = "blue")
legend(
  "bottom",
  inset = c(0, -.45),
  legend = legend.idents,
  text.width = 2.5,
  ncol = 2,
  pch = c(13, 14),
  col = c("red", "blue")
)
```

Running the prior code gives the following results[1] based on our assumptions and logic.

```
          Mean NPV of decision [$M]
Decision1                 2.39
Decision2                 2.91
[1] "The difference in mean value between strategies: $ 0.524 M"
```

---

[1]The values you observe could vary due to the simulation error that arises from R's simple Monte Carlo engine. Simple Monte Carlo can be somewhat "noisy."

The chart of the cumulative probability is shown in Figure 15-1.

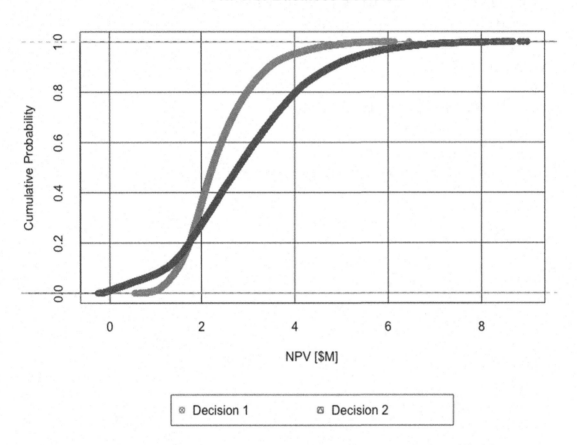

Figure 15-1. *The overlapping cumulative distribution functions (CDFs) for Decisions 1 and 2 illustrate intervals of dominance of one decision compared to another*

Decreasing dominance of one decision over another implies that we face increasing ambiguity about which decision to choose clearly. In such cases, we need to know which uncertainties need more attention for determining how to choose clearly. VOI analysis can tell us how much to spend on that attention.

The results in Figure 15-1 show that, given our current state of knowledge about our potential investment opportunity decision choices, we face a dilemma. This dilemma derives from classic finance theory, which tells that, if we are rational, we should prefer those investment opportunities with the highest average (or mean) return and those with the least variance. Here we face the situation in which the best decision based on average return (Decision 2) is inferior to Decision 1 based on its overall variance (Decision 2 is nearly twice as wide as Decision 1 in its full range of potential outcome). Furthermore, we observe that Decision 2 neither strictly dominates over Decision 1 nor stochastically dominates it.

Strict Dominance   There is no sample from the highest valued decision that is lower than any sample from a lower valued decision. Looking at both the probability density function (PDF) and cumulative distribution function (CDF) curves (Figure 15-2), the position of the lowest tail of the highest valued decision would at least be just separated from the highest tail of the lower valued decision.

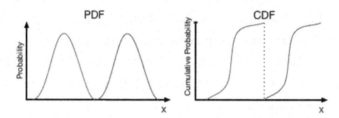

Figure 15-2. *Strict dominance illustrated by the spatial relationship of probability distributions*

Stochastic Dominance   The tails of the PDF curves overlap, but CDF curves do not cross over as the highest valued decision remains strictly offset from the lower valued decision across the full range of variation (Figure 15-3).

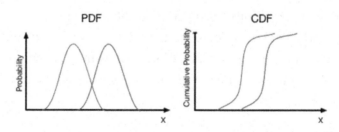

Figure 15-3. *Stochastic dominance illustrated by the spatial relationship of probability distributions*

In fact, we observe that the lower tail of Decision 2 crosses over and extends beyond the lower tail of Decision 1 to expose us potentially to a lower outcome than had we chosen Decision 1. The situation we face by our analysis is ambiguous, clearly implying that before we commit to a specific decision pathway, we might want to refine our current state of knowledge. Which of the pieces of information should we focus on? We can develop guidance on how to improve our current state of knowledge with a kind of two-way sensitivity analysis that I have alluded to previously, the tornado analysis.

The Sensitivity Analysis

What we need is a way to prioritize our attention on the uncertain variables based on our current quality of information about them and how strongly the likely range of their behavior might affect the average value of the objective function, NPV. Because we know that the NPV curves overlap, one or more of the uncertainties are causing this overlap as they are the only sources of variation in the model. So, before we go seeking higher quality information about our uncertainties, we first need to understand which of the probable range of outcomes for any of the uncertainties is significant enough to potentially make us regret choosing our initial inclination to accept Decision 2. To accomplish this, we use tornado sensitivity analysis.

Tornado sensitivity analysis works by sequentially observing how much the average NPV changes in response to the 80th percentile range of each uncertain variable. We choose a variable and set it to its p10 value, then we record the effect on average NPV. Next we set the same variable to its p90 and record the effect on average NPV. During both of these iterations, we can set the other uncertainties to their mean value or let them run according their defined variation. Whereas the former approach is generally faster, the latter approach is more accurate, especially if the business model represents a highly nonlinear transform of the inputs to the objective function. If the business model used exhibits strong nonlinearity, Jensens's inequality[2] effects can manifest.[3] The code that I provide here follows the more accurate latter approach. Repeating this process for each variable, we observe how much each variable influences the objective function both by its functional strength and across the range of its assessed likelihood of occurrence.

---

[2]E.K. Godunova (originator), "Jensen inequality." In *Encyclopedia of Mathematics*. http://www.encyclopediaofmath.org/index.php?title=Jensen_inequality&oldid=16975

[3]Jensen's inequality, stated as $E(f(X))) \geq f(E(X))$, demonstrates that it is not necessarily the case that the expected value of a function of descriptor variables **X** is equal to the function of the expected values of its descriptor variables. The equality holds true for linear systems of equations, but usually not for nonlinear ones. Our model has a nonlinearity in it due to the inclusion of the time value of money effects on the cash flow. The skewness in many of the underlying uncertainties compounds the problem of this distortion.

Before we look at the sensitivity analysis code, let's recall the following characteristics about each of the variables in our model, both the kind of information they represent and their units.

- time -> the time index, year

- peak units sold -> uncertainty

- time to peak units sold -> uncertainty, years

- price -> uncertainty, $/unit

- sales, general, and admin -> uncertainty, % revenue

- cost of goods sold -> uncertainty, $/unit

- tax rate -> % profit

- initial investment -> uncertainty, $

- discount rate -> %/year

The following code (which I've named Sensitivity_Analysis.R) requires that the namespace and sample values of the variables that come from Business_Decision_Model.R be available for use in R's workspace; therefore, make sure you run Business_Decision_Model.R before running this code.

```
# Create a vector of values representing the sensitivity quantiles.
sens.range <- c(0.1, 0.5, 0.9)

npv.sens1 <- CalcModelSensitivity(
  unc.list = dec1.uncs,
  sens.q = sens.range
) / 1e6

npv.sens2 <- CalcModelSensitivity(
  unc.list = dec2.uncs,
  sens.q = sens.range
) / 1e6

print("Decision 1 Sensitivity to Uncertainty [$M]")
print(signif(npv.sens1[, c(1, length(sens.range))], 3))

print("Decision 2 Sensitivity to Uncertainty [$M]")
print(signif(npv.sens2[, c(1, length(sens.range))], 3))
```

The results of the sensitivity analysis look like the following tables.

```
[1] "Decision 1 Sensitivity to Uncertainty [$M]"
                0.1  0.9
peak.units.sold 1.72 3.21
time.to.peak    2.69 2.03
price           2.24 2.51
cogs            2.48 2.29
sga             2.46 2.31
investment      2.41 2.36

[1] "Decision 2 Sensitivity to Uncertainty [$M]"
                0.1  0.9
peak.units.sold 1.44 4.35
time.to.peak    3.50 2.39
price           2.74 3.06
cogs            3.02 2.79
sga             3.04 2.78
investment      2.94 2.87
```

Note that the `peak.units.sold` variable for Decision 1 returns a value of $1.72 million under the 0.1 column. This is the average NPV of Decision 1 when `peak.units.sold` is set to its p10 value (e.g., 10,000) while all the other uncertain variables follow their natural variation. Likewise, it returns an average NPV = $3.21 million when it is set to its p90 value (e.g., 18,000) while all the other uncertain variables follow their natural variation. The rest of the table is read in a similar manner.

Please note that a random seed of 98 is used in the Assumptions. (`set.seed(98)`). Keeping this value set to any fixed value will ensure that you always get the same values between simulation runs of the model. If you change this value, you will observe slightly different values in the reported tables. If you remove the `set.seed(98)` statement altogether (or comment it out), you will see different values every time you run the model. Just how stable these values remain between runs indicates how sensitive your model is to the noisiness of simple Monte Carlo simulation. Therefore, it is often helpful to run a model several times to get a good feel for this stability (or lack of it), then select a seed value that does a good job of reflecting the ensemble of tests. Otherwise, when you report your values to others who aren't familiar with the nuances of Monte Carlo simulation, you will face having to explain the nuances of Monte Carlo simulation. Guess how productive that conversation usually is.

Because we already know that Decision 2 has a higher average NPV than Decision 1, we need to know which uncertainties could potentially cause us regret for taking that higher average valued decision. We answer that question by observing which uncertainties cause overlap in the range of NPV between the decisions. If we look down the rows of uncertainties, we see that the two uncertainties that cause such an overlap of value are the `peak.units.sold` and `time.to.peak`, as the lowest value of Decision 2 for each of these variables overlaps the highest one for Decision 1.

We refer to these overlapping uncertainties as *critical uncertainties* because they pose the greatest potential for making us wish we had taken a different route once we commit to the execution of a decision. This is not to say that the other uncertainties are not important and will not need to be monitored once we do commit to a decision; rather, it simply means that for the purpose of making a clear decision now, these critical uncertainties are the greatest contributors to our current state of ambiguity.

Normally, this list of uncertainties should be graphed as a vertical floating bar chart such that the following are the case.

- Each bar represents the range of variation caused by an uncertainty around the mean value of a given decision.

- The order of the uncertainties follows a declining order of importance as determined by the range between the `p10` and `p90` of each variable.

Overlaying both of these charts provides a better visual cue about which uncertainties should be deemed critical. Here is the code I use to produce the iconic tornado charts for each decision to obtain that visual insight. You can append this to the end of the `Sensitivity_Analysis.R` file.

```
# Find the rank order the uncertainties by the declining range of variation
# they cause in the NPV. Base this on the decision with the highest mean
value.
mean.bdm.list <- c(m.bdm1, m.bdm2)
bdm.sensitivity.list <- list(npv.sens1, npv.sens2)
npv.sens.rank <-
  order(abs(bdm.sensitivity.list[[which.max(mean.bdm.list)]][, 1] -
          bdm.sensitivity.list[[which.max(mean.bdm.list)]][,
          length(sens.range)]),
      decreasing = FALSE)
```

```r
# Reorder the variables in the sensitivity arrays and names array by the rank
# order.
ranked.npv.sens2 <- npv.sens2[npv.sens.rank, c(1, length(sens.range))]
ranked.npv.sens1 <- npv.sens1[npv.sens.rank, c(1, length(sens.range))]

# Plot the tornado chart for Decision 2
par(mai = c(1, 1.75, .5, .5))
barplot(
  t(ranked.npv.sens2) - m.bdm2,
  main = "NPV Sensitivity to Uncertainty Ranges",
  names.arg = names(ranked.npv.sens2),
  col = "blue",
  xlim = c(min(npv.sens1, npv.sens2), max(npv.sens1, npv.sens2)),
  xlab = "Decision2 NPV [$M]",
  beside = TRUE,
  horiz = TRUE,
  offset = m.bdm2,
  las = 1,
  space = c(-1, 1),
  cex.names = 1
)

# Plot the tornado chart for Decision 1
par(mai = c(1, 1.75, .5, .5))
barplot(
  t(ranked.npv.sens1) - m.bdm1,
  main = "NPV Sensitivity to Uncertainty Ranges",
  names.arg = names(ranked.npv.sens1),
  col = "red",
  xlim = c(min(npv.sens1, npv.sens2), max(npv.sens1, npv.sens2)),
  xlab = "Decision1 NPV [$M]",
  beside = TRUE,
  horiz = TRUE,
```

```
    offset = m.bdm1,
    las = 1,
    space = c(-1, 1),
    cex.names = 1
)
```

Note that we use the rank ordering for Decision 2 as the basis for ordering the display of uncertainty effects across decisions. This will keep the uncertainties across decisions on the same row if it's the case that the rank ordering differs between decisions. Notice also that we keep the colors consistent with the original CDF charts, and we set the xlimits (xlim = c(min(npv.sens1, npv.sens2), max(npv.sens1, npv.sens2))) for both graphs to comprehend the full range of sensitivity across all the decisions. This latter chart parameter setting ensures that both charts are scaled across the same relative range and that the relative width of the bars for each chart remains consistent.

In the tornado charts shown in Figure 15-4, overlapping bars across decisions reveal critical uncertainties that are candidates for VOI analysis. In this case, peak.units.sold and time.to.peak present themselves as the critical uncertainties.

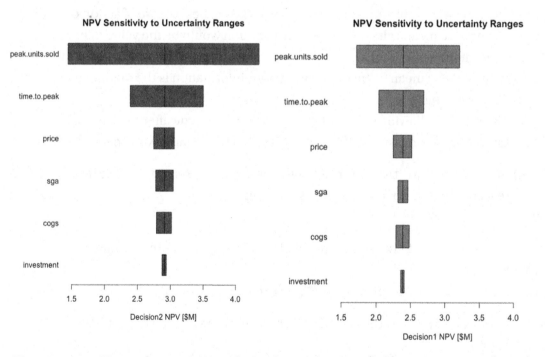

Figure 15-4. *Tornado sensitivity charts for Decision 2 (left) vs. Decision 1 (right)*

VOI Algorithms

Following the pattern that I established from the beginning of this discussion, we develop the code for VOI using a simple three-branch decision tree. The purpose is to reinforce our understanding of the method. However, because this method can mask over subtleties in the resultant distributions, we finish the discussion by developing the code that goes much further to preserve those details.

Coarse Focus First

Recall that the CDF curves for the decision values we produced when we first ran `Business_Decision_Model.R` demonstrated the effect of all the uncertainties acting on the NPV for each decision. Now that we know which uncertainties need our attention for VOI analysis, we need to isolate their effects from those of the other uncertainties. This is easy enough to do now because we already did this in Chapter 14 by computing the reversed decision tree with an analog matrix calculation. Effectively what we're doing by this is acting as if we have prior knowledge about which outcome will occur more to our favor in many possible future states, and we take the best course of action on each iterated state. The mean value of this new distribution would be the value of having perfect information prior to making a decision. The difference between this value and the highest decision mean value without the prior information is the rational maximum we should be willing to pay for that prior knowledge, the VOI.

Again, run the following `Value_of_Information_1.R` code after the `Business_Decision_Model.R` and `Sensitivity_Analysis.R` to retain the R workspace values.

```
# Define a vector of the extended Swanson-Megill probability weights. These
# correspond to the values in the sens.range vector.
prob.wts <- c(0.3, 0.4, 0.3)

# Create the 3x3 matrix of the products of the uncertainty branch
probabilities.
probs.branch.matrix <- prob.wts %*% t(prob.wts)

# Typically, we run business model for each decision with critical
uncertainty
# peak.units.sold set to its respective p10, p50, p90 values. Find the mean
NPV
```

```r
# of the business model at each of these points. However, we already did this in
# the Sensitivity_Analysis.R script, so all we need to do is use the slice of
# the sensitivity table associated with the peak.units.sold row.
bdm1.sens <- npv.sens1[1, ]
bdm2.sens <- npv.sens2[1, ]

# Create a 3x3 matrix of the values in the bdm1.sens and bdm2.sens vectors.
# Transpose the values in the second matrix
bdm1.sens.matrix <- array(rep(bdm1.sens, 3), dim = c(3, 3))
bdm2.sens.matrix <- t(array(rep(bdm2.sens, 3), dim = c(3, 3)))

# Find the parallel maximum value between these matrices. This represents
# knowing the maximum value given the prior information about the combinations
# of the outcomes of each uncertainty.
bdm.sens.prior.info <- pmax(bdm1.sens.matrix, bdm2.sens.matrix)

# Find the expected value of the matrix that contains the decision values with
# prior information. This is the value of knowing the outcome before making a
# decision.
bdm.prior.info <- sum(bdm.sens.prior.info * probs.branch.matrix)

# Find the maximum expected decison value without prior information.
bdm.max.curr.info <- max(mean(bdm1), mean(bdm2))

# The value of information is the net value of knowing
# the outcome beforehand compared to the decison with
# the highest value before the outcome is known.
val.info <- (bdm.prior.info - bdm.max.curr.info)

print("Decision Value [$M]")
value.report <- list("Prior Information" = signif(bdm.prior.info, 3),
                     "Current Information" = signif(bdm.max.curr.info, 3),
                     "Value of Information" = signif(val.info, 3))
```

```
print(value.report)
```

The final results for the VOI analysis produce the following report.

```
[1] "Decision Value [$M]"
$`Prior Information`
[1] 3.2

$`Current Information`
[1] 2.91

$`Value of Information`
[1] 0.291
```

This implies that if we could buy perfect information about the outcome of peak. units.sold prior to making a decision, we should be willing to spend no more than approximately $300,000 to gain that insight. As we will soon discuss, this value is a coarse focus result. We might be able to improve this value some with a fine focus approach of including more granularity in our uncertainty branches.

We can also interpret VOI in a slightly different manner. Imagine that we are poised at just the point in time before we commit to a decision. In some sense two potential universes exist ahead of each of our decisions, and each universe evolves along multiple potential alternate routes depending on how each conditional uncertainty manifests itself. Suppose that each route has a coordinate pair associated with it indicating the decision and route such that the first possible route for Decision 1 would be $(d, i) = (1, 1)$, and the first possible route for Decision 2 would be $(d, i) = (2, 1)$. The next possible set of routes would be $(d, i) = (1, 2)$ and $(d, i) = (2, 2)$, and so forth for each iteration of our simulated universes. A clairvoyant's crystal ball makes it possible for us to compare routes $(1, i)$ and $(2, i)$ simultaneously for free with the added benefit of observing only the variation on the NPV of each decision due to the peak.units.sold. Even though Decision 2 has a greater average NPV than Decision 1, there will be some future routes in which the NPV of Decision 1 exceeds that of Decision 2. Given that we prefer futures with higher values and that we are (hopefully) rational, we will always choose the future route (d, i) with the highest value. As a consequence, the resulting value distribution of chosen routes will manifest a higher average NPV than Decision 2. The average incremental improvement in value across those rationally chosen future routes would equal approximately $300,000.

The Finer Focus

The approach I have followed so far relies on the extended Swanson-Megill probability weights to approximate the probabilities of the outcomes of the p10, p50, p90 branches of the decision tree. As approximations, these weights do not strictly conform to the required probability weights of any arbitrary distribution such that its mean and variance are perfectly preserved. The extended Swanson-Megill weights are "rules of thumb" weights for estimating the mean of a distribution, not exact predictors of it.

Of course, the only way to have an exact predictor of an uncertain event's expected value is to have all the possible samples associated with it (which could be impossible) or to have a symmetry or structure about the event that dictates the expected value, as in the case with fair coins, dice, or a deck of cards. However, we could improve the precision of our calculation simply by using more sensitivity points that represent discrete bins along the distribution of the critical uncertainty instead of predefined quantiles. The bins are analogs of decision tree pathways just as the quantile points were. Given that, we can replicate our matrix approach, but in a more general fashion.

Once our sensitivity analysis identifies a critical uncertainty (i.e., `peak.units.sold`), we can use it for the more granular analysis. We start by finding the histogram of the critical uncertainty for each decision. (If you do not want to plot the histograms, set the `plot` parameter to `FALSE`.)

```
# Find the histogram of the critical uncertainty for each decision pathway.
dec1.branch.hist <- hist(dec1.uncs$peak.units.sold, plot = TRUE)
dec2.branch.hist <- hist(dec2.uncs$peak.units.sold, plot = TRUE)
```

From these histograms, we need appropriate values from the critical uncertainty to serve as test values to find the contingent NPV values. The plot of the histogram is based on the breakpoints in the domain of the uncertainty, but we don't need the breakpoints. We need values that represent the bins that are demarcated by the breakpoints. Fortunately, R provides this information in the calculation of the histogram and stores it in the $mid list element.

```
# Find the midpoints of the bins.
dec1.branch.bins <- dec1.branch.hist$mids
dec2.branch.bins <- dec2.branch.hist$mids
```

The histogram returns the following values for the Decision 1 and 2 branch bins:

```
> dec1.branch.bins
 [1]  6500  7500  8500  9500 10500 11500 12500 13500 14500 15500 16500 17500
[13] 18500 19500 20500 21500 22500 23500 24500 25500 26500
```

```
> dec2.branch.bins
 [1]  1000  3000  5000  7000  9000 11000 13000 15000 17000 19000 21000 23000
[13] 25000 27000 29000 31000 33000 35000 37000
```

Note that the first set of bins is longer than the second set.

Next, we need to find a vector of the frequency of the bins in the histograms. We accomplish this by using the $counts values for each bin, then dividing them by the number of simulation samples.

```
1  dec1.branch.cnts <- dec1.branch.hist$counts
2  dec1.branch.probs <- dec1.branch.cnts / samps
3
4  dec2.branch.cnts <- dec2.branch.hist$counts
5  dec2.branch.probs <- dec2.branch.cnts / samps
```

We observe the following frequency values for each bin value.

```
> dec1.branch.probs
 [1] 0.01893333 0.02743333 0.02530000 0.02740000 0.15906667 0.15916667
 [7] 0.11796667 0.07133333 0.07136667 0.07366667 0.07326667 0.07350000
[13] 0.01326667 0.01260000 0.01213333 0.01353333 0.01183333 0.01096667
[19] 0.01193333 0.01233333 0.00300000
```

```
> dec2.branch.probs
 [1] 0.02296667 0.02220000 0.02163333 0.02183333 0.05960000 0.10106667
 [7] 0.09976667 0.10246667 0.10070000 0.10110000 0.09603333 0.10083333
[13] 0.05650000 0.01743333 0.01683333 0.01673333 0.01703333 0.01636667
[19] 0.00890000
```

Each element in the respective vector represents the probability that the given bin, a branch in the decision tree, will manifest; therefore, each of these vectors should sum to 1.

Now we can find the sensitivity of the decision expected NPV by iterating across the uncertainty bins, using each bin as a value for peak.units.sold in the CalcBizModel() function. Make sure to replace the peak.units.sold parameter with a vector of the current bin with a length equal to the sample size (i.e., dec1.uncs$peak.units.sold -> rep(dec1.branch.bins[b], samps)).

```
# Find the sensitivity of the NPV to the midpoints of the bins in each
# uncertainty.
bdm1.sens <- c(0)
for (b in 1:length(dec1.branch.bins)) {
  bdm1.sens[b] <- mean(
    CalcBizModel(
      time,
      samps,
      rep(dec1.branch.bins[b], samps),
      dec1.uncs$time.to.peak,
      dec1.uncs$price,
      dec1.uncs$sga,
      dec1.uncs$cogs,
      tax.rate,
      dec1.uncs$investment,
      disc.rate
    )$npv / 1e6
  )
}

bdm2.sens <- c(0)
for (b in 1:length(dec2.branch.bins)) {
  bdm2.sens[b] <- mean(
    CalcBizModel(
```

```
    time,
    samps,
    rep(dec2.branch.bins[b], samps),
    dec2.uncs$time.to.peak,
    dec2.uncs$price,
    dec2.uncs$sga,
    dec2.uncs$cogs,
    tax.rate,
    dec2.uncs$investment,
    disc.rate
  )$npv / 1e6
 )
}
```

We observe the following results:

```
> bdm1.sens
[1] 1.072199 1.255956 1.439714 1.623472 1.807229 1.990987 2.174744 2.358502
[9] 2.542260 2.726017 2.909775 3.093533 3.277290 3.461048 3.644805 3.828563
[17] 4.012321 4.196078 4.379836 4.563594 4.747351
```

```
> bdm2.sens
[1] 0.01501496 0.37361250 0.73221004 1.09080758 1.44940511 1.80800265
[7] 2.16660019 2.52519773 2.88379527 3.24239281 3.60099034 3.95958788
[13] 4.31818542 4.67678296 5.03538050 5.39397804 5.75257558 6.11117311
[19] 6.46977065
```

Now we follow the pattern in the simpler example. Create the MxN matrix of the Cartesian products of the uncertainty branch probabilities.

```
probs.branch.matrix <- dec1.branch.probs %*% t(dec2.branch.probs)
```

Figure 15-5 graphically illustrates this last step.

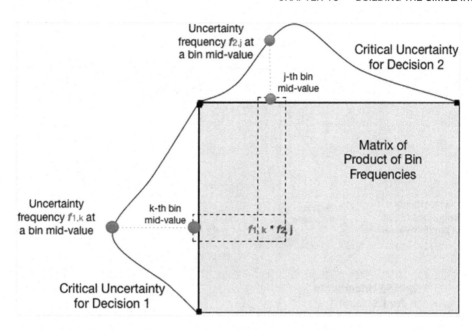

Figure 15-5. *The matrix of the product of the bin frequencies of the critical uncertainty under current inspection is the Cartesian product of those bin frequencies*

Next, create an MxN matrix of the values in the bdm1.sens and bdm2.sens vectors. Transpose the values in the second matrix.

```
bdm1.sens.matrix <-
   array(rep(bdm1.sens, length(dec2.branch.probs)),
       dim = c(length(dec1.branch.probs), length(dec2.branch.probs)))
```

```
bdm2.sens.matrix <-
   t(array(rep(bdm2.sens, length(dec1.branch.probs)),
       dim = c(length(dec2.branch.probs), length(dec1.branch.probs)
)))
```

Figure 15-6 demonstrates the parallel relationship of the results of finding the contingent relationship of the decision value to each critical uncertainty's bins' midvalues.

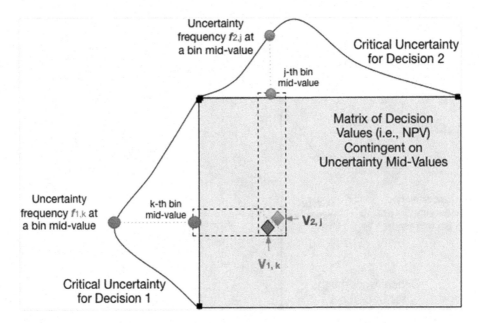

Figure 15-6. *The matrix of the decision values of each possible branch combination that occurs on the outcome of the critical uncertainties' midvalues*

Figure 15-7 illustrates finding the parallel maximum value between these matrices. Recall that this represents knowing the maximum value given the prior information about the combinations of the outcomes of each uncertainty.

```
bdm.sens.prior.info <- pmax(bdm1.sens.matrix, bdm2.sens.matrix)
```

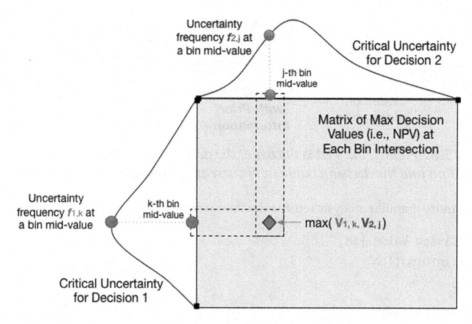

Figure 15-7. *The decision value of each possible branch outcome is found by the pairwise (or parallel) max of each possible outcome combination*

Find the expected value of the matrix that contains the decision values with prior information (Figure 15-8). This is the value of knowing the outcome before making a decision.

```
bdm.prior.info <- sum(bdm.sens.prior.info * probs.branch.matrix)
```

$$\begin{matrix} \textit{Decision Value} \\ \textit{with Prior} \\ \textit{Information} \end{matrix} = \sum_{k,j} \max(V_{1,k}, V_{2,j}) * f_{1,k} * f_{2,j}$$

Figure 15-8. *The steps outlined in Figures 15-5 through 15-7 are summarized by this expression*

Find the maximum expected decison value without prior information.

```
bdm.max.curr.info <- max(mean(bdm1), mean(bdm2))
```

The VOI is the net value of knowing the outcome beforehand compared to the decison with the highest value before the outcome is known (Figure 15-9).

```
val.info <- (bdm.prior.info - bdm.max.curr.info)
```

$$\begin{array}{c}\textbf{\textit{Value of}}\\ \textbf{\textit{Information}}\end{array} = \begin{array}{c}\textbf{\textit{Decision Value}}\\ \textbf{\textit{with Prior}}\\ \textbf{\textit{Information}}\end{array} - \begin{array}{c}\textbf{\textit{Decision Value}}\\ \textbf{\textit{with Current}}\\ \textbf{\textit{Information}}\end{array}$$

$$\begin{array}{c}\textbf{\textit{Value of}}\\ \textbf{\textit{Information}}\end{array} = \begin{array}{c}\textbf{\textit{Decision Value}}\\ \textbf{\textit{with Prior}}\\ \textbf{\textit{Information}}\end{array} - \text{max}(V_1, V_2)$$

Figure 15-9. *Finally, the VOI is the net of the decision value with prior information and the decision value with current information*

This more granular analysis returns the following report:

```
[1] "Decision Value [$M]"
$`Prior Information`
[1] 3.28

$`Current Information`
[1] 2.91

$`Value of Information`
[1] 0.372
```

As in the previous sections, I placed the raw script code for this section in a file called Value_of_Information_2.R. This file should also be run after the Sensitivity_Analysis.R script.

Notice that Prior Information and Value of Information values are a little higher than those obtained by using our coarse focus extended Swanson-Megill branches. Of course, this is due to the fact that our fine-focus use of more branches extends into the tails of the critical uncertainties (which might extend much farther than the truncated inner 80th percentile prediction interval), and we are using a finer distinction in the detail of the whole domain of the critical uncertainty than the ham-fisted discrete blocks of the extended Swanson-Megill.

As you might have inferred, I liken the VOI analysis process to that of using an optical microscope to focus attention on interesting details. Using the extended Swanson-Megill branches is perfectly fine for initially identifying the critical uncertainties as one would use the coarse focus knob. The final approach described here is like using the fine focus knob to gain insights from a more granular inspection of the distribution of probability.

Please note that the approach described in this tutorial applies to calculating VOI on continuous uncertainties by providing a means to discretize the uncertainty into a manageable and informative set of branches. If the critical uncertainty is discrete to begin with, there's no need to go through the discretization process. One would just use the assessed probabilities for the specific discrete branches.

Is `histogram()` the Best Way to Find the Uncertainty Bins?   In this last section, I chose to use the `histogram()` function to discretize the critical uncertainty. This approach is not necessary. In fact, you could use your own defined set of breakpoints that span the full range of uncertain variable behavior, maybe with block sizes of your own choosing or block sizes determined by equally spaced probability. Then you would find the associated frequencies and midpoints within these blocks. Although that might give you a sense of accomplishment of going through the process of writing the code, let me suggest that although VOI is important, it is not a value that needs to be pursued to an extremely high degree of precision. Its purpose is to point us in the right direction to resolve decision ambiguity, to limit the unnecessary expenditure of resources to resolve that ambiguity, and to ensure that whatever resources we do apply, they are applied in a resource-efficient manner. The `histogram()` function finds an appropriate discretization well enough.

To ensure that the order of operations taken through this chapter is clear to reproduce the code and its results, you can follow the flowchart in Figure 15-10. Appendix E contains the full uninterrupted source code according to this order, too.

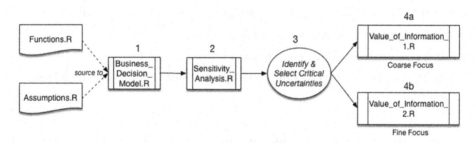

Figure 15-10. *The steps outlined in this chapter to calculate value of information according to the source code files that are used*

Concluding Comments

The world is full of uncertainties that frustrate our ability to choose clearly. If we think long enough about what those might be, it doesn't take long to be overwhelmed. The problem is that we live in an economic universe, a place where there are an unlimited number of needs and desires, but only a limited amount of available resources to address those needs. Finding the right information of sufficient quality to clarify the impact of uncertainties is no less an economic concern than determining the right allocation of capital in a portfolio to achieve desired returns.

An interesting irony has arisen in the information age: We are swamped in data, yet we struggle to comprehend its information content. We might think that having access to all the data we now have would significantly reduce the anxiety of choosing clearly. To be sure, advances in data science and Big Data management have produced some great insights for some organizations that have invested in those capabilities, but whether investments in data science and Big Data programs are mostly economically productive remains an open question.[4] The problem, it seems, is that data still need to be parsed, cleaned, and tested for their contextual relationship to the strategic questions at hand before systematic relationships between inputs and outputs are understood well enough to reduce uncertainty about which related decisions are valuable. Not only are we concerned about which uncertainties matter, we now have to ask which data matter, and it isn't always clear from the beginning of such efforts that if we apply resources to understand which data matter that we will, in fact, arrive at valuable understandings. Regardless, the most important activity any kind of analyst can do to improve the quality of his or her efforts is to frame the analytic problem well before cleaning any data, getting more data, or settling on the best analysis and programming environment. Indeed, solving the right problem accurately is much more important than solving irrelevant problems with high precision or technical sophistication. The advent of the age of Big Data has, perhaps, compounded our original problem.

Every day, I enjoy three shots of espresso. I hope the summarizing shots that follow are just as enjoyable to you in your quest to improve your business case analysis skill set.

---

[4]See, for example, Eric Almquist, John Senior, and Tom Springer, "Three Promises and Perils of Big Data," Bain Brief, April 8, 2015: "Through 2017, 60% of Big Data projects will fail to go beyond piloting and experimentation and will be abandoned." Although I can't prove it here, I suspect that most of these failures occur for the same reasons all projects fail, the most frequent and pernicious one being the failure to properly frame the reason for the project anyway.

Espresso Shot 1

Evaluating VOI helps us address the issues of living in an economic universe. VOI focuses and concentrates our attention on the issues that matter, like an espresso of information. As you noticed in the tornado charts we developed, the width of the bars displays a type of Pareto distribution; that is, the amount of variation observed in the objective measure that is attributable to any uncertainty appears to decline in an exponential manner. The effect demonstrated here did not derive from cleverly chosen values to emphasize a point. This pattern repeats itself regularly. Personally, I've observed over dozens and dozens of decision analysis efforts that included anywhere from 10 to 100 uncertainties that the largest amount of uncertainty in the objective is attributable to between 10% and 20% of the uncertainties in question. So here's an important understanding: Not all uncertainties we face are equally important to the level of worry we initially lend them. The twist of lemon peel is that what we were originally biased to focus on the most actually matters the least.

Espresso Shot 2

The tornado sensitivity analysis delivers a bonus feature. When we compare the charts between important decisions, we see that not all of the most significant uncertainties really matter either. For the purpose of choosing clearly, not every significant uncertainty is critical. Again, from my experience, of the 10% to 20% of uncertainties that are significant, generally no more than one or two are critical.

Espresso Shot 3

The first two shots of information espresso should significantly reduce our worry and anxiety about what can cause us harm and regret as well as reduce the amount of activity we spend trying to obtain better information about them. Now we have a third shot of espresso: When we have identified the critical uncertainties, we can know just how much we should rationally budget to get that higher quality information. VOI analysis, properly applied, should save us worry, time, and money, making us more economically efficient and competitive with the use of our limited resources. As my grandfather used to say, "Don't spend all of that in one place." When you are called to spend it, though, don't spend more than you need to. VOI tells us what that no-more-than-you-need-to actually is.

Deterministic Model

You can download the source code from this book's product page at
www.apress.com/9781484234945 by clicking the source code button there. If you want to
read this source code in a noncomputing medium (i.e., ebook or print book), Appendices A
through E also contain the uninterrupted R source code and SME elicitation instructions.

The deterministic model as discussed in Chapter 2 is maintained in a file directory
structure in the R application working directory that looks like the following:

```
~/BizSimWithR
      Determ_Model.R
      Determ_Model_Sensitivity.R
      /libraries
            My_Functions.R
      /data
            global_assumptions.R
            risk_assumptions.csv
```

Note that the global_assumptions.R and risk_assumptions.csv files are used by
both the deterministic and risk-based models.

The following is the contents of the global_assumptions.R file.

```
# Global Model Assumptions
kHorizon <- 20 # yrs, length of time the model covers.
year <- 1:kHorizon # an index for temporal calculations.
kSampsize <- 1000 # the number of iterations in the Monte Carlo simulation.
run <- 1:kSampsize # the iteration index.
kTaxRate <- 38 # %
kDiscountRate <- 12 # %/yr, used for discounted cash flow calculations.
kDeprPer <- 7 # yrs, the depreciation schedule for the capital.
```

© Robert D. Brown III 2018
R. D. Brown III, *Business Case Analysis with R*, https://doi.org/10.1007/978-1-4842-3495-2

The variable data collected in the CSV file called `risk_assumptions.csv` looks like this:

◇	A	B	C	D	E	F
1	variable	p10	p50	p90	units	notes
2	p1.capex	35000000	40000000	47000000	$	phase 1 capital spend to be spread over the duration of Phase 1.
3	p1.dur	1	2	4	yrs	phase 1 duration.
4	p2.capex	15000000	20000000	30000000	$	phase 2 capital spend to be spread over the duration of Phase 1.
5	p2.dur	1	2	4	yrs	phase 2 duration.
6	maint.capex	1500000	2000000	3000000	$	annual maintenance spend.
7	fixed.prod.cost	2250000	3000000	4500000	$	annual fixed production costs.
8	prod.cost.escal	2.25	3	4.5	%/yr	the rate at which the fixed costs grow.
9	var.prod.cost	2.625	3.5	5.25	$/lb	the variable production costs.
10	var.cost.redux	3.75	5	7.5	%/yr	the rate at which the variable production costs decay.
11	gsa.rate	19	20	22	% revenue	general, sales & administration cost.
12	time.to.peak.sales	3.75	5	7.5	yrs	the number of years to reach max sales.
13	mkt.demand	3.75	5	7.5	ktons/yr	the amount of annual sales at peak.
14	price	4.5	6	9	$/lb	product price.
15	rr.comes.to.market	0	0.6	1		the probability that RoadRunner comes to market.
16	rr.time.to.market	4	5	6	yrs	the time elapsed until RoadRunner comes to market.
17	early.market.share	45	50	65	%	the market share available to Acme if RoadRunner comes to market early.
18	late.market.share	70	75	85	%	the market share available to Acme if RoadRunner comes to market late.
19	price.redux	12	15	20	%	reduction in price that that occurs if RoadRunner comes to market.

The R source code that drives the initial deterministic simulation follows here.

```
# Read source data and function files. Modify the path names to match your
# directory structure and file names.
source("~/BizSimWithR/data/global_assumptions.R")
d.data <- read.csv("~/BizSimWithR/data/risk_assumptions.csv")
source("~/BizSimWithR/libraries/My_Functions.R")

# Slice the values from data frame d.data.
d.vals <- d.data$p50[1:13]

# Assign p50 values to variables.
p1.capex <- d.vals[1]
p1.dur <- d.vals[2]
p2.capex <- d.vals[3]
p2.dur <- d.vals[4]
maint.capex <- d.vals[5]
fixed.prod.cost <- d.vals[6]
prod.cost.escal <- d.vals[7]
var.prod.cost <- d.vals[8]
var.cost.redux <- d.vals[9]
gsa.rate <- d.vals[10]
time.to.peak.sales <- d.vals[11]
```

```
mkt.demand <- d.vals[12]
price <- d.vals[13]

# CAPEX Module
phase <- (year <= p1.dur) * 1 +
  (year > p1.dur & year <= (p1.dur + p2.dur)) * 2 +
  (year > (p1.dur + p2.dur)) * 3

capex <- (phase == 1) * p1.capex / p1.dur +
  (phase == 2) * p2.capex / p2.dur +
  (phase == 3) * maint.capex

# Depreciation Module
depr.matrix <-
  t(sapply(year, function(y)
    ifelse(
      y <= p1.dur & year > 0,
      0,
      ifelse(
        y == (p1.dur + 1) & year < y + kDeprPer & year >= y,
        p1.capex / kDeprPer,
        ifelse((year >= y) & (year < (y + kDeprPer)),
               capex[y - 1] / kDeprPer, 0)
      )
    ))
  )
depr <- colSums(depr.matrix)

# Sales Module
mkt.adoption <- pmin(cumsum(phase > 1) / time.to.peak.sales, 1)
sales <- mkt.adoption * mkt.demand * 1000 * 2000
revenue <- sales * price

# OPEX Module
fixed.cost <- (phase > 1) * fixed.prod.cost *
  (1 + prod.cost.escal / 100) ^ (year - p1.dur - 1)

var.cost <- var.prod.cost * (1 - var.cost.redux / 100) ^
  (year - p1.dur - 1) * sales
```

```r
gsa <- (gsa.rate / 100) * revenue
opex <- fixed.cost + var.cost

# Value
gross.profit <- revenue - gsa
op.profit.before.tax <- gross.profit - opex - depr
tax <- op.profit.before.tax * kTaxRate / 100
op.profit.after.tax <- op.profit.before.tax - tax
cash.flow <- op.profit.after.tax + depr - capex
cum.cash.flow <- cumsum(cash.flow)

discount.factors <- 1 / (1 + kDiscountRate / 100) ^ year
# Following the convention for when payments are counted as occurring
# at the end of a time period.
discounted.cash.flow <- cash.flow * discount.factors
npv <- sum(discounted.cash.flow)

# Pro Forma
# Create an array of the variables to be used in the pro forma.
pro.forma.vars <- array(
  c(
    sales,
    revenue,
    -gsa,
    gross.profit,-fixed.cost,
    -var.cost,
    -opex,
    -depr,
    op.profit.before.tax,
    -tax,
    op.profit.after.tax,
    depr,
    -capex,
    cash.flow
  ),
  dim = c(kHorizon, 14)
)
```

```
# Create a data frame of the pro forma array.
pro.forma <- data.frame(pro.forma.vars)

# Assign text names to a vector. These will be the column headers of the
# data frame.
pro.forma.headers <-
  c(
    "Sales [lbs]",
    "Revenue",
    "GS&A",
    "Gross Profit",
    "Fixed Cost",
    "Variable Cost",
    "OPEX",
    "-Depreciation",
    "Operating Profit Before Tax",
    "Tax",
    "Operating Profit After Tax",
    "+Depreciation",
    "CAPEX",
    "Cash Flow"
  )

# Coerces the default column headers to be the headers we like.
colnames(pro.forma) <- pro.forma.headers
rownames(pro.forma) <- year

# Transposes the data frame so that columns become rows.
pro.forma = t(pro.forma)
```

The R source code that drives the sensitivity analysis of the NPV follows here.

```
# Read source data and function files. Modify the path names to match your
# directory structure and file names.
source("/Applications/R/RProjects/BizSimWithR/data/global_assumptions.R")
d.data <-
read.csv("/Applications/R/RProjects/BizSimWithR/data/risk_assumptions.csv")
source("/Applications/R/RProjects/BizSimWithR/libraries/My_Functions.R")
```

```r
# Slice the values from data frame d.data.
d.vals <- d.data$p50[1:13]

d.vals.sens.points <- d.data[1:13, 2:4]
sens.point <- 1:3
len.d.vals <- length(d.vals)
len.sens.range <- length(sens.point)
npv.sens <- array(0, c(len.d.vals, len.sens.range))

for (i in 1:len.d.vals) {
  for (k in 1:len.sens.range) {
    d.vals[i] <-  d.vals.sens.points[i, k]

    # Assign values to variables.
    p1.capex <- d.vals[1]
    p1.dur <- d.vals[2]
    p2.capex <- d.vals[3]
    p2.dur <- d.vals[4]
    maint.capex <- d.vals[5]
    fixed.prod.cost <- d.vals[6]
    prod.cost.escal <- d.vals[7]
    var.prod.cost <- d.vals[8]
    var.cost.redux <- d.vals[9]
    gsa.rate <- d.vals[10]
    time.to.peak.sales <- d.vals[11]
    mkt.demand <- d.vals[12]
    price <- d.vals[13]

    # CAPEX Module
    phase <- (year <= p1.dur) * 1 +
      (year > p1.dur & year <= (p1.dur + p2.dur)) * 2 +
      (year > (p1.dur + p2.dur)) * 3

    capex <- (phase == 1) * p1.capex / p1.dur +
      (phase == 2) * p2.capex / p2.dur +
      (phase == 3) * maint.capex
```

```r
# Depreciation Module
depr.matrix <-
  t(sapply(year, function(y)
    ifelse(
      y <= p1.dur & year > 0,
      0,
      ifelse(
        y == (p1.dur + 1) & year < y + kDeprPer & year >= y,
        p1.capex / kDeprPer,
        ifelse((year >= y) & (year < (y + kDeprPer)),
               capex[y - 1] / kDeprPer, 0)
      )
    ))))
depr <- colSums(depr.matrix)

# Sales Module
mkt.adoption <- pmin(cumsum(phase > 1) / time.to.peak.sales, 1)
sales <- mkt.adoption * mkt.demand * 1000 * 2000
revenue <- sales * price

# OPEX Module
fixed.cost <- (phase > 1) * fixed.prod.cost *
  (1 + prod.cost.escal / 100) ^ (year - p1.dur - 1)

var.cost <- var.prod.cost * (1 - var.cost.redux / 100) ^
  (year - p1.dur - 1) * sales

gsa <- (gsa.rate / 100) * revenue
opex <- fixed.cost + var.cost

# Value
gross.profit <- revenue - gsa
op.profit.before.tax <- gross.profit - opex - depr
tax <- op.profit.before.tax * kTaxRate / 100
op.profit.after.tax <- op.profit.before.tax - tax
cash.flow <- op.profit.after.tax + depr - capex
cum.cash.flow <- cumsum(cash.flow)
```

```
    discount.factors <- 1 / (1 + kDiscountRate / 100) ^ year
    # Following the convention for when payments are counted as occurring
    # at the end of a time period.
    discounted.cash.flow <- cash.flow * discount.factors
    npv <- sum(discounted.cash.flow)
    npv.sens[i, k] <- npv
  }
  d.vals[i] <- d.data$p50[i]
}

# Assign npv.sens to a data frame.
var.names <- d.data[1:13, 1]
sens.point.names <- c("p10", "p50", "p90")
rownames(npv.sens) <- var.names
colnames(npv.sens) <- sens.point.names

# Sets up the sensitivity array.
npv.sens.array <- array(0, c(len.d.vals, 2))
npv.sens.array[, 1] <- (npv.sens[, 1] - npv.sens[, 2])
npv.sens.array[, 2] <- (npv.sens[, 3] - npv.sens[, 2])
rownames(npv.sens.array) <- var.names
colnames(npv.sens.array) <- sens.point.names[-2]

# Calculates the rank order of the NPV sensitivity based on the
# absolute range caused by a given variable. The npv.sens.array
# is reindexed by this rank ordering for the bar plot.
npv.sens.rank <-
  order(abs(npv.sens.array[, 1] - npv.sens.array[, 2]),
        decreasing = FALSE)

ranked.npv.sens.array <- npv.sens.array[npv.sens.rank,]
ranked.var.names <- var.names[npv.sens.rank]
rownames(ranked.npv.sens.array) <- ranked.var.names

# Plots the sensitivity array.
par(mai = c(1, 1.75, 0.5, 0.5))
barplot(
  t(ranked.npv.sens.array) / 1000,
```

```
  main = "Deterministic NPV
  Sensitivity",
  names.arg = ranked.var.names,
  col = "light blue",
  xlab = "NPV [$000]",
  beside = TRUE,
  horiz = TRUE,
  offset = npv.sens[, 2] / 1000,
  las = 1,
  space = c(-1, 1),
  cex.names = 1,
  tck = 1
)
```

APPENDIX B

Risk Model

The risk model as discussed in Chapters 3 and 4 is maintained in a file directory structure in the R application working directory that looks like the following:

```
~/BizSimWithR
      Risk_Model.R
      Risk_Model_Sensitivity.R
      /libraries
            My_Functions.R
      /data
            global_assumptions.R
            risk_assumptions.csv
```

Recall that the global_assumptions.R and risk_assumptions.csv files are used by both the deterministic and risk-based models. Because their content is shown in Appendix A, it is not repeated here.

The R source code that drives the risk simulation follows here.

```
# Read source data and function files. Modify the path names to match your
# directory structure and file names.
source("~/BizSimWithR/data/global_assumptions.R")
d.data <- read.csv("~/BizSimWithR/data/risk_assumptions.csv")
source("~/BizSimWithR/libraries/My_Functions.R")

# Slice the values from data frame d.data.
d.vals <- d.data[, 2:4]

# Assign values to variables using appropriate distributions.
p1.capex <- CalcBrownJohnson(0, d.vals[1, 1], d.vals[1, 2],
                             d.vals[1, 3], kSampsize)
```

© Robert D. Brown III 2018
R. D. Brown III, *Business Case Analysis with R*, https://doi.org/10.1007/978-1-4842-3495-2

```
p1.dur <- round(CalcBrownJohnson(1, d.vals[2, 1], d.vals[2, 2],
                             d.vals[2, 3], kSampsize), 0)
p2.capex <- CalcBrownJohnson(0, d.vals[3, 1], d.vals[3, 2],
                             d.vals[3, 3], kSampsize)
p2.dur <- round(CalcBrownJohnson(1, d.vals[4, 1], d.vals[4, 2],
                             d.vals[4, 3], kSampsize), 0)
maint.capex <- CalcBrownJohnson(0, d.vals[5, 1], d.vals[5, 2],
                             d.vals[5, 3], kSampsize)
fixed.prod.cost <- CalcBrownJohnson(0, d.vals[6, 1], d.vals[6, 2],
                               d.vals[6, 3], kSampsize)
prod.cost.escal <- CalcBrownJohnson(, d.vals[7, 1], d.vals[7, 2],
                               d.vals[7, 3], kSampsize)
var.prod.cost <- CalcBrownJohnson(0, d.vals[8, 1], d.vals[8, 2],
                               d.vals[8, 3], kSampsize)
var.cost.redux <- CalcBrownJohnson(, d.vals[9, 1], d.vals[9, 2],
                               d.vals[9, 3], kSampsize)
gsa.rate <- CalcBrownJohnson(0, d.vals[10, 1], d.vals[10, 2],
                            d.vals[10, 3], 100, kSampsize)
time.to.peak.sales <- round(CalcBrownJohnson(1, d.vals[11, 1],
                                        d.vals[11, 2], d.vals[11, 3],
                                        kSampsize), 0)
mkt.demand <- CalcBrownJohnson(0, d.vals[12, 1], d.vals[12, 2],
                            d.vals[12, 3], kSampsize)
price <- CalcBrownJohnson(0, d.vals[13, 1], d.vals[13, 2],
                        d.vals[13, 3], kSampsize)
rr.comes.to.market <- rbinom(kSampsize, 1, d.vals[14, 2])
rr.time.to.market <- round(CalcBrownJohnson(1, d.vals[15, 1],
                                       d.vals[15, 2], d.vals[15, 3],
                                       kSampsize), 0)
early.market.share <- CalcBrownJohnson(0, d.vals[16, 1],
                                   d.vals[16, 2], d.vals[16, 3], 100,
                                   kSampsize)
late.market.share <- CalcBrownJohnson(0, d.vals[17, 1],
                                  d.vals[17, 2], d.vals[17, 3], 100,
                                  kSampsize)
```

```
price.redux <- CalcBrownJohnson(0, d.vals[18, 1], d.vals[18, 2],
                                d.vals[18, 3], kSampsize)

# CAPEX module
phase <- t(sapply(run, function(r)
  (year <= p1.dur[r]) * 1 +
    (year > p1.dur[r] & year <= (p1.dur[r] + p2.dur[r])) * 2 +
    (year > (p1.dur[r] + p2.dur[r])) * 3))

capex <-
  t(sapply(run, function(r)
    (phase[r,] == 1) * p1.capex[r] / p1.dur[r] +
      (phase[r,] == 2) * p2.capex[r] / p2.dur[r] +
      (phase[r,] == 3) * maint.capex[r]))

# Depreciation module
depr.matrix <-
  array(sapply(run, function(r)
    sapply(year, function(y)
      ifelse(
        y <= p1.dur[r] & year > 0,
        0,
        ifelse(
          y == (p1.dur[r] + 1) &
            year < y + kDeprPer & year >= y,
          p1.capex[r] / kDeprPer,
          ifelse((year >= y) &
                   (year < (y + kDeprPer)), capex[r, y - 1] / kDeprPer, 0)
        )
      ))),
    dim = c(kHorizon, kHorizon, kSampsize))

depr <- t(sapply(run, function(r)
  sapply(year, function(y)
    sum(depr.matrix[y, r]))))

# Competition module
market.share <-
```

```
  (rr.comes.to.market == 1) * ((rr.time.to.market <= p1.dur) *
                              early.market.share / 100 + (rr.time.
                              to.market > p1.dur) *
                              late.market.share / 100
  ) +
  (rr.comes.to.market == 0) * 1
# Sales module
mkt.adoption <- t(sapply(run, function(r)
  market.share[r] *
    pmin(cumsum(phase[r,] > 1) / time.to.peak.sales[r], 1)))
sales <-
  t(sapply(run, function(r)
    mkt.adoption[r,] * mkt.demand[r] *
      1000 * 2000))
revenue <- t(sapply(run, function(r)
  sales[r,] * price[r] *
    (1 - rr.comes.to.market[r] * price.redux[r] / 100)))

# OPEX module
fixed.cost <-
  t(sapply(run, function(r)
    (phase[r,] > 1) * fixed.prod.cost[r] *
      (1 + prod.cost.escal[r] / 100) ^ (year - p1.dur[r] - 1)))
var.cost <- t(sapply(run, function(r)
  var.prod.cost[r] *
    (1 - var.cost.redux[r] / 100) ^ (year - p1.dur[r] - 1) * sales[r,]))
gsa <- t(sapply(run, function(r)
  (gsa.rate[r] / 100) * revenue[r,]))
opex <- fixed.cost + var.cost

# Value
gross.profit <- revenue - gsa
op.profit.before.tax <- gross.profit - opex - depr
tax <- op.profit.before.tax * kTaxRate / 100
op.profit.after.tax <- op.profit.before.tax - tax
cash.flow <- op.profit.after.tax + depr - capex
```

```
cum.cash.flow <- t(sapply(run, function(r)
  cumsum(cash.flow[r,])))

# Following the convention for when payments are counted as occurring
# at the end of a time period.
discount.factors <- 1 / (1 + kDiscountRate / 100) ^ year
discounted.cash.flow <- t(sapply(run, function(r)
  cash.flow[r,] *
    discount.factors))
npv <- sapply(run, function(r)
  sum(discounted.cash.flow[r,]))
mean.npv <- mean(npv)

# Calculates the 80th percentile quantiles in the cash flow and
# cumulative cash flow.
q80 <- c(0.1, 0.5, 0.9)
cash.flow.q80 <-
  sapply(year, function(y)
    quantile(cash.flow[, y], q80))
cum.cash.flow.q80 <-
  sapply(year, function(y)
    quantile(cum.cash.flow[, y],
             q80))

# Plots the 80th percentile cash flow quantiles.
plot(
  0,
  type = "n",
  xlim = c(1, kHorizon),
  ylim = c(min(cash.flow.q80) / 1000,
           max(cash.flow.q80) / 1000),
  xlab = "Year",
  ylab = "[$000]",
  main = "Cash Flow",
  tck = 1
)
```

```
lines(
  year,
  cash.flow.q80[1, ] / 1000,
  type = "b",
  lty = 1,
  col = "blue",
  pch = 16
)
lines(
  year,
  cash.flow.q80[2,] / 1000,
  type = "b",
  lty = 1,
  col = "red",
  pch = 18
)
lines(
  year,
  cash.flow.q80[3,] / 1000,
  type = "b",
  lty = 1,
  col = "darkgreen",
  pch = 16
)
legend(
  "topleft",
  legend = q80,
  bg = "grey",
  pch = c(16, 18, 16),
  col = c("blue", "red", "dark green")
)

# Plots the 80th percentile cumulative cash flow quantiles.
plot(
  0,
  type = "n",
```

```
  xlim = c(1, kHorizon),
  ylim = c(min(cum.cash.flow.q80) / 1000,
           max(cum.cash.flow.q80) / 1000),
  xlab = "Year",
  ylab = "[$000]",
  main = "Cumulative Cash Flow",
  tck = 1
)
lines(
  year,
  cum.cash.flow.q80[1, ] / 1000,
  type = "b",
  lty = 1,
  col = "blue",
  pch = 16
)
lines(
  year,
  cum.cash.flow.q80[2, ] / 1000,
  type = "b",
  lty = 1,
  col = "red",
  pch = 18
)
lines(
  year,
  cum.cash.flow.q80[3, ] / 1000,
  type = "b",
  lty = 1,
  col = "darkgreen",
  pch = 16
)
```

```r
legend(
  "topleft",
  legend = q80,
  bg = "grey",
  pch = c(16, 18, 16),
  col =
    c("blue", "red", "dark green")
)

# Calculates and plots the histogram of NPV.
breakpoints <-
  seq(min(npv), max(npv), abs(min(npv) - max(npv)) / 20)
hist(
  npv / 1000,
  freq = FALSE,
  breaks = breakpoints / 1000,
  main = "Histogram of NPV",
  xlab = "NPV [$000]",
  ylab = "Probability Density",
  col = "blue"
)

# Calculates the cumulative NPV probability chart and table.
cum.quantiles <- seq(0, 1, by = 0.05)
cum.npv.vals <-
  quantile(npv, cum.quantiles) #plot these values in chart
cum.npv.frame <- data.frame(cum.npv.vals) #table

# Plot the cumulative probability NPV curve.
plot(
  cum.npv.vals / 1000,
  cum.quantiles,
  main = "Cumulative Probability of NPV",
  xlab = "NPV [$000]",
  ylab = "Cumulative Probability",
  "b",
  tck = 1,
```

```r
  col = "blue",
  pch = 16
)

# Pro forma
# Create a data frame of the variables' mean values to be used in the
# pro forma.
pro.forma.vars <- array(
  c(
    sales,
    revenue,
    -gsa,
    gross.profit,-fixed.cost,
    -var.cost,
    -opex,
    -depr,
    op.profit.before.tax,
    -tax,
    op.profit.after.tax,
    depr,
    -capex,
    cash.flow
  ),
  dim = c(kSampsize, kHorizon, 14)
)

# Finds the annual mean of each pro forma element.
mean.pro.forma.vars <- array(0, c(14, kHorizon))

for (p in 1:14) {
  mean.pro.forma.vars[p,] <- sapply(year, function(y)
    mean(pro.forma.vars[, y, p]))
}

pro.forma <- data.frame(mean.pro.forma.vars)
```

```
# Assign text names to a vector. These will be the column headers of
# the data frame.
pro.forma.headers <-
  c(
    "Sales [lbs]",
    "Revenue",
    "GS&A",
    "Gross Profit",
    "Fixed Cost",
    "Variable Cost",
    "OPEX",
    "-Depreciation",
    "Operating Profit Before Tax",
    "Tax",
    "Operating Profit After Tax",
    "+Depreciation",
    "CAPEX",
    "Cash Flow"
  )

# Coerces the default column headers to be the headers we like.
colnames(pro.forma) <- year
rownames(pro.forma) <- pro.forma.headers

# Waterfall chart
# Extract the rows from the pro forma for the waterfall chart.
waterfall.rows <- c(2, 3, 5, 6, 10, 13)
waterfall.headers <- pro.forma.headers[waterfall.rows]
wf.pro.forma <- pro.forma[waterfall.rows,]

# Find the present value of the extracted pro forma elements.
pv.wf.pro.forma <- rep(0, length(waterfall.rows))
for (w in 1:length(waterfall.rows)) {
  pv.wf.pro.forma[w] <- sum(wf.pro.forma[w,] * discount.factors)
}
```

```
cum.pv.wf.pro.forma1 <- cumsum(pv.wf.pro.forma)
cum.pv.wf.pro.forma2 <- c(0, cum.pv.wf.pro.forma1[1:(length(waterfall.rows)
-1)])
wf.high <- pmax(cum.pv.wf.pro.forma1, cum.pv.wf.pro.forma2)
wf.low <- pmin(cum.pv.wf.pro.forma1, cum.pv.wf.pro.forma2)

waterfall <- array(0, c(2, length(waterfall.rows)))
waterfall[1,] <- wf.high
waterfall[2,] <- wf.low
colnames(waterfall) <- waterfall.headers
rownames(waterfall) <- c("high", "low")

# Plot the waterfall.
boxplot(
  waterfall / 1000,
  data = waterfall / 1000,
  notch = FALSE,
  main = "Waterfall Chart",
  xlab = waterfall.headers,
  ylab = "$000",
  col = c("blue", rep("red", 5))
)
```

The R source code that drives the sensitivity analysis of the NPV from the risk model follows here.

```
# Read source data and function files. Modify the path names to match your
# directory structure and file names.
source("~/BizSimWithR/data/global_assumptions.R")
d.data <-
  read.csv("~/BizSimWithR/data/risk_assumptions.csv")
source("~/BizSimWithR/libraries/My_Functions.R")

# Slice the values from data frame d.data.
d.vals <- d.data[, 2:4]

sens.range <- c(0.1, 0.9)
```

```
len.d.vals <- length(d.vals[, 1])
len.sens.range <- length(sens.range)
npv.sens <- array(0, c(len.d.vals, len.sens.range))

# Assign values to variables using appropriate distributions.
p1.capex <- CalcBrownJohnson(0, d.vals[1, 1], d.vals[1, 2],
                              d.vals[1, 3], kSampsize)
p1.dur <- round(CalcBrownJohnson(1, d.vals[2, 1], d.vals[2, 2],
                                d.vals[2, 3], kSampsize), 0)
p2.capex <- CalcBrownJohnson(0, d.vals[3, 1], d.vals[3, 2],
                              d.vals[3, 3], kSampsize)
p2.dur <- round(CalcBrownJohnson(1, d.vals[4, 1], d.vals[4, 2],
                                d.vals[4, 3], kSampsize), 0)
maint.capex <- CalcBrownJohnson(0, d.vals[5, 1], d.vals[5, 2],
                                d.vals[5, 3], kSampsize)
fixed.prod.cost <- CalcBrownJohnson(0, d.vals[6, 1], d.vals[6, 2],
                                    d.vals[6, 3], kSampsize)
prod.cost.escal <- CalcBrownJohnson(, d.vals[7, 1], d.vals[7, 2],
                                    d.vals[7, 3], kSampsize)
var.prod.cost <- CalcBrownJohnson(0, d.vals[8, 1], d.vals[8, 2],
                                  d.vals[8, 3], kSampsize)
var.cost.redux <- CalcBrownJohnson(, d.vals[9, 1], d.vals[9, 2],
                                   d.vals[9, 3], kSampsize)
gsa.rate <- CalcBrownJohnson(0, d.vals[10, 1], d.vals[10, 2],
                             d.vals[10, 3], 100, kSampsize)
time.to.peak.sales <- round(CalcBrownJohnson(1, d.vals[11, 1],
                                        d.vals[11, 2], d.vals[11, 3],
                                        kSampsize), 0)
mkt.demand <- CalcBrownJohnson(0, d.vals[12, 1], d.vals[12, 2],
                               d.vals[12, 3], kSampsize)
price <- CalcBrownJohnson(0, d.vals[13, 1], d.vals[13, 2],
                          d.vals[13, 3], kSampsize)
rr.comes.to.market <- rbinom(kSampsize, 1, d.vals[14, 2])
rr.time.to.market <- round(CalcBrownJohnson(1, d.vals[15, 1],
                                        d.vals[15, 2], d.vals[15, 3],
                                        kSampsize), 0)
```

```
early.market.share <- CalcBrownJohnson(0, d.vals[16, 1],
                                       d.vals[16, 2], d.vals[16, 3], 100,
                                       kSampsize)
late.market.share <- CalcBrownJohnson(0, d.vals[17, 1],
                                      d.vals[17, 2], d.vals[17, 3], 100,
                                      kSampsize)
price.redux <- CalcBrownJohnson(0, d.vals[18, 1], d.vals[18, 2],
                                d.vals[18, 3], kSampsize)

d.vals.vect <- c(
  p1.capex,
  p1.dur,
  p2.capex,
  p2.dur,
  maint.capex,
  fixed.prod.cost,
  prod.cost.escal,
  var.prod.cost,
  var.cost.redux,
  gsa.rate,
  time.to.peak.sales,
  mkt.demand,
  price,
  rr.comes.to.market,
  rr.time.to.market,
  early.market.share,
  late.market.share,
  price.redux
)

d.vals.temp <- array(d.vals.vect, dim = c(kSampsize, len.d.vals))
d.vals.temp2 <- d.vals.temp
d.vals2 <- d.vals[-2]

CalcBizSim = function(x) {
  # x is the data array that contains the presimulated samples for
  # each variable.
```

```
p1.capex <- x[, 1]
p1.dur <- x[, 2]
p2.capex <- x[, 3]
p2.dur <- x[, 4]
maint.capex <- x[, 5]
fixed.prod.cost <- x[, 6]
prod.cost.escal <- x[, 7]
var.prod.cost <- x[, 8]
var.cost.redux <- x[, 9]
gsa.rate <- x[, 10]
time.to.peak.sales <- x[, 11]
mkt.demand <- x[, 12]
price <- x[, 13]
rr.comes.to.market <- x[, 14]
rr.time.to.market <- x[, 15]
early.market.share <- x[, 16]
late.market.share <- x[, 17]
price.redux <- x[, 18]

# CAPEX module
phase <- t(sapply(run, function(r)
  (year <= p1.dur[r]) * 1 +
    (year > p1.dur[r] & year <= (p1.dur[r] + p2.dur[r])) * 2 +
    (year > (p1.dur[r] + p2.dur[r])) * 3))

capex <-
  t(sapply(run, function(r)
    (phase[r,] == 1) * p1.capex[r] / p1.dur[r] +
      (phase[r,] == 2) * p2.capex[r] / p2.dur[r] +
      (phase[r,] == 3) * maint.capex[r]))

# Depreciation module
depr.matrix <-
  array(sapply(run, function(r)
    sapply(year, function(y)
      ifelse(
        y <= p1.dur[r] & year > 0,
        0,
```

```
      ifelse(
        y == (p1.dur[r] + 1) &
          year < y + kDeprPer & year >= y,
        p1.capex[r] / kDeprPer,
        ifelse((year >= y) &
                  (year < (y + kDeprPer)), capex[r, y - 1] / kDeprPer, 0)
      )
    ))),
  dim = c(kHorizon, kHorizon, kSampsize))

depr <- t(sapply(run, function(r)
  sapply(year, function(y)
    sum(depr.matrix[y, r])))))

# Competition module
market.share <-
  (rr.comes.to.market == 1) * ((rr.time.to.market <= p1.dur) *
                                early.market.share / 100 + (rr.time.
                                to.market > p1.dur) *
                                late.market.share / 100
  ) +
  (rr.comes.to.market == 0) * 1

# Sales module
mkt.adoption <- t(sapply(run, function(r)
  market.share[r] *
    pmin(cumsum(phase[r,] > 1) / time.to.peak.sales[r], 1)))
sales <-
  t(sapply(run, function(r)
    mkt.adoption[r,] * mkt.demand[r] *
      1000 * 2000))
revenue <- t(sapply(run, function(r)
  sales[r,] * price[r] *
    (1 - rr.comes.to.market[r] * price.redux[r] / 100)))
```

```
  # OPEX module
  fixed.cost <-
    t(sapply(run, function(r)
      (phase[r,] > 1) * fixed.prod.cost[r] *
        (1 + prod.cost.escal[r] / 100) ^ (year - p1.dur[r] - 1)))
  var.cost <- t(sapply(run, function(r)
    var.prod.cost[r] *
      (1 - var.cost.redux[r] / 100) ^ (year - p1.dur[r] - 1) * sales[r,]))
  gsa <- t(sapply(run, function(r)
    (gsa.rate[r] / 100) * revenue[r,]))
  opex <- fixed.cost + var.cost

  # Value
  gross.profit <- revenue - gsa
  op.profit.before.tax <- gross.profit - opex - depr
  tax <- op.profit.before.tax * kTaxRate / 100
  op.profit.after.tax <- op.profit.before.tax - tax
  cash.flow <- op.profit.after.tax + depr - capex
  cum.cash.flow <- t(sapply(run, function(r)
    cumsum(cash.flow[r,])))

  # Following the convention for when payments are counted as occurring
  # at the end of a time period.
  discount.factors <- 1 / (1 + kDiscountRate / 100) ^ year
  discounted.cash.flow <- t(sapply(run, function(r)
    cash.flow[r,] *
      discount.factors))
  npv <- sapply(run, function(r)
    sum(discounted.cash.flow[r,]))
  return(npv)
}

base.mean <- mean(CalcBizSim(d.vals.temp))

for (i in 1:len.d.vals) {
  for (k in 1:len.sens.range) {
    # For a given variable, replace its samples with a vector containing
```

```
  # each sensitivity endpoint.
  d.vals.temp2[, i] <-  rep(d.vals2[i, k], kSampsize)

  # Calculate the mean NPV by calling the CalcBizSim() function.
  mean.npv <- mean(CalcBizSim(d.vals.temp2))

  # Insert the resultant mean NPV into an array that catalogs the
  # variation in the mean NPV by each variable's sensitivity points.
  npv.sens[i, k] <- mean.npv
 }

 # Restore the current variable's last sensitivity point with its original
 # simulated samples.
 d.vals.temp2[, i] <- d.vals.temp[, i]
}

# Assign npv.sens to a data frame.
var.names <- d.data$variable
rownames(npv.sens) <- d.data$variable
colnames(npv.sens) <- sens.range

# Sets up the sensitivity array.
npv.sens.array <- array(0, c(len.d.vals, 2))
npv.sens.array[, 1] <- (npv.sens[, 1] - base.mean)
npv.sens.array[, 2] <- (npv.sens[, 2] - base.mean)
rownames(npv.sens.array) <- var.names
colnames(npv.sens.array) <- sens.range

# Calculates the rank order of the NPV sensitivity based on the
# absolute range caused by a given variable. The npv.sens.array
# is reindexed by this rank ordering for the bar plot.
npv.sens.rank <- order(abs(npv.sens.array[, 1] -
                           npv.sens.array[, 2]), decreasing = FALSE)
ranked.npv.sens.array <- npv.sens.array[npv.sens.rank,]
ranked.var.names <- var.names[npv.sens.rank]
rownames(ranked.npv.sens.array) <- ranked.var.names
```

```
# Plots the sensitivity array.
par(mai = c(1, 1.75, 0.5, 0.5))
barplot(
  t(ranked.npv.sens.array) / 1000,
  main = "NPV Sensitivity to
  Uncertainty Ranges",
  names.arg = ranked.var.names,
  col = "red",
  xlab = "NPV [$000]",
  beside = TRUE,
  horiz = TRUE,
  offset = base.mean / 1000,
  las = 1,
  space = c(-1, 1),
  cex.names = 1
)
```

Simulation and Finance Functions

I keep functions that I develop in a file called MyFunctions.R, which I import at the beginning of every R project. The following represents the BrownJohnson distribution function and my finance function set.

```
# BrownJohnson distribution
CalcBrownJohnson = function(minlim=-Inf, p10, p50, p90, maxlim=Inf, n,
samponly=TRUE) {
# This function simulates a distribution from three quantile
# estimates for the probability intervals of the predicted outcome.
# p10 = the 10th percentile estimate
# p50 = the 50th percentile estimate
# p90 = the 90th percentile estimate
# n = the number of runs used in the simulation
# Note, this function strictly requires that p10 < p50 < p90.
# The process of simulation is simple Monte Carlo.
# The returned result is, by default, a vector of values for X if
# samponly=TRUE, else a (n x 2) matrix is returned such that the first
# column contains the domain samples in X, and the second column
# contains the uniform variate samples from U.

  if (p10 == p50 && p50 == p90) {
            return(rep(p50, n))
      } else if (p10 >= p50 || p50 >= p90) {
    stop("Parameters not given in the correct order. Must be
```

© Robert D. Brown III 2018
R. D. Brown III, *Business Case Analysis with R*, https://doi.org/10.1007/978-1-4842-3495-2

```
    given as p10 < p50 < p90.")
  } else {
#Create a uniform variate sample space in the interval (0,1).
    U <- runif(n, 0, 1)

# Calculates the virtual tails of the distribution given the p10, p50, p90
# inputs. Truncates the tails at the upper and lower limiting constraints.
    p0 <- max(minlim, 2.5 * p10 - 1.5 * p50)
    p100 <- min(maxlim, 2.5 * p90 - 1.5 * p50)

#This next section finds the linear coefficients of the system of linear
# equations that describe the linear spline, using linear algebra.
# [C](A) = (X)
# (A) = [C]^-1 * (X)
# In this case, the elements of (C) are found using the values (0, 0.1,
# 0.5, 0.9, 1) at the endpoints of each spline segment. The elements
# of (X) correspond to the values of (p0, p10, p10, p50, p50, p90,
# p90, p100). Solving for this system of linear equations gives linear
# coefficients that transform values in U to intermediate values in X.
# Because there are four segments in the linear spline, and each
# segment contains two unknowns, a total of eight equations are
# required to solve the system.

# The spline knot values in the X domain.
    knot.vector <- c(p0, p10, p10, p50, p50, p90, p90, p100)

# The solutions to the eight equations at the knot points required to
# describe the linear system.
    coeff.vals <- c(0, 1, 0, 0, 0, 0, 0, 0, 0.1, 1, 0, 0, 0, 0, 0, 0,
    0, 0, 0.1, 1, 0, 0, 0, 0, 0, 0, 0.5, 1, 0, 0, 0, 0, 0, 0, 0, 0, 0.5,
    1, 0, 0, 0, 0, 0, 0.9, 1, 0, 0, 0, 0, 0, 0, 0, 0, 0.9, 1, 0, 0, 0,
    0, 0, 0, 1, 1)

# The coefficient matrix created from the prior vector looks like the
# following matrix:
#    [, 1]    [, 2]    [, 3]    [, 4]    [, 5]    [, 6]    [, 7]    [, 8]
# [1,] 0.0     1.0      0.0      0.0      0.0      0.0      0.0      0.0
# [2,] 1.0     1.0      0.0      0.0      0.0      0.0      0.0      0.0
```

```
# [3,] 0.0     0.0     1.0     1.0     0.0     0.0     0.0     0.0
# [4,] 0.0     0.0     0.5     1.0     0.0     0.0     0.0     0.0
# [5,] 0.0     0.0     0.0     0.0     0.5     1.0     0.0     0.0
# [6,] 0.0     0.0     0.0     0.0     0.9     1.0     0.0     0.0
# [7,] 0.0     0.0     0.0     0.0     0.0     0.0     0.9     1.0
# [8,] 0.0     0.0     0.0     0.0     0.0     0.0     1.0     1.0

    coeff.matrix <- t(matrix(coeff.vals, nrow=8, ncol=8))


# The inverse of the coefficient matrix.
    inv.coeff.matrix <- solve(coeff.matrix)

# The solution vector of the linear coefficients.
    sol.vect <- inv.coeff.matrix %*% knot.vector

#Builds the response by the piecewise linear sections
    X <- (U <= 0.1) * (sol.vect[1, 1] * U + sol.vect[2, 1]) +
      (U > 0.1 & U <= 0.5) * (sol.vect[3, 1] * U + sol.vect[4, 1]) +
      (U > 0.5 & U <= 0.9) * (sol.vect[5, 1] * U + sol.vect[6, 1]) +
      (U > 0.9 & U <= 1) * (sol.vect[7, 1] * U + sol.vect[8, 1])

    if (samponly == TRUE) {
      return(X)
    } else {
      X <- array( c(X, U), dim=c(n, 2))
      return(X)
    }

# Plots the array automatically after calculation.
# Comment out as necessary.
# plot(X[, 1], X[, 2], type="p", xlab="Outcomes", ylab="Cumulative
# Probability")
  }
}

# NPV function
CalcNPV = function(series, time, dr, eotp = TRUE) {
  # series = the cash flow series
  # time = index over which series is allocated
```

251

```r
  # dr = discount rate
  # eotp = end of time period calculation (default is TRUE)
  # or beginning of time period calculation (FALSE)

  this.NPV <- sum(series / (1 + dr) ^ (time - (1 - eotp)))
}

# IRR function
CalcIRR = function (series,
                    time,
                    irr0 = 0.1,
                    tolerance = 0.00001) {
  #Calculates the discount rate that produces a 0 NPV
  # of a series of end-of-year cash flows.
  # series = a vector of cash flows
  # time = the time array index
  # irr0 = the initial guess
  # tolerance = the error around 0 at which the goal seek stops

  irr <- c(irr0)
  npv <- CalcNPV(series, time, irr[1])
  i <- 1
  while (abs(npv[i]) > tolerance) {
    if (i == 1) {
      if (npv[i] > 0) {
        (irr[i + 1] = irr[i] * 2)
      } else {
        (irr[i + 1] <- irr[i] / 2)
      }
    } else {
      # uses Newton-Raphson method of convergence
      slope.npv <- (npv[i] - npv[i - 1]) /
        (irr[i] - irr[i - 1])
      irr[i + 1] <- irr[i] - npv[i] / slope.npv
    }
```

```
    npv[i + 1] <- CalcNPV(series, time, irr[i + 1])
    i <- i + 1
  }
  irr <- irr[i]
}

# Equivalent Period Interest Rate function
CalcEqPerIR = function(i, N) {
  # Calculates the equivalent interest rate for subperiod of a stated
    period
  # interest rate
  # i = the stated period interest rate
  # N = the number of subperiods within the stated period interest rate

  eq.per.ir = ((1 + i) ^ (1 / N)) - 1
}

# Period Payment function
CalcPerPayment = function(i, L, N) {
  # Calculates the period payment due on a loan amount, L, to be paid back at
  # the number of payment periods specified, N, at the period interest
  # rate i.

  per.payment = L * i / (1 - (1 + i) ^ (-N))
}

# Period Principal Payment function
CalcPerPrinPayment = function(i, L, M, N) {
  # Returns the period N principal payments due on a loan L obtained at N=0.
  # The loan has a life of M periods paid back at a period interest rate i.
  # N should be a sequence from 1 to M.

  per.pr.in.payment = L * i * (1 / (1 - (1 + i) ^ (-M)) - 1) * (1 + i) ^
    (N - 1)
}
```

```
# Period Interest Payment function
CalcPerInPayment = function(i, L, M, N) {
  # Returns the period N interest payments due on a loan L obtained at N=0.
  # The loan has a life of M periods paid back at a period interest rate i.
  # N should be a sequence from 1 to M.

  per.in.payment = l * i * ((1 + i) ^ (N - 1) - ((1 + i) ^ (N - 1) - 1) /
                                  (1 - (1 + i) ^ (-M)))
}

# Period Remaining Balance function
CalcPerRemBal = function(i, L, M, N) {
  # Returns the period N remaining balance due on a loan L obtained at N=0.
  # The loan has a life of M periods paid back at a period interest rate i.
  # N should be a sequence from 1 to M.

  per.rem.bal = L * ((1 + i) ^ N - ((1 + i) ^ N - 1) / (1 - (1 + i) ^ (-M)))
}

# Future Worth of an Annuity function
CalcFutureWorthAnnuity = function(A, i, N) {
  # The future worth of an annuity for a given period of time is the sum at
  # the end of that period of the future worths of all payments at a given
  # rate of interest each compounding period.
  # A = the amount of the annuity
  # i = the period interest rate
  # N = the period at the end of the compounding

  future.worth.annuity = (A * (1 + i) ^ N - 1) / i
}

# Present Worth of an Annuity function
CalcPrsWorthAnnuity = function(A, i, N) {
  # The present worth of an annuity for a given period of time is the sum at
  # the start of that period of the present worths of each payment in the
  # series, at a given rate of interest each compounding period.
  # interest each compounding period.
```

```
  # A = the amount of the annuity
  # i = the period interest rate
  # N = the period at the end of the compounding

  prs.worth.annuity = A * ((1 + i) ^ N - 1) / (i * (1 + i) ^ N)
}

# Annuity from a Present Amount function
CalcAnnuityPrsAmount = function(P, i, N) {
  # The annuity from a present amount is the period amount that can be
  # withdrawn for a definite period of time from a present sum of money
  # at a given rate of interest each compounding period.
  # P = the present value
  # i = the period interest rate
  # N = the period at which the present amount is depleted

  annuity.present.amount = P * i * (1 + i) ^ N / ((1 + i) ^ N - 1)
}

# Weighted Average Cost of Capital function
CalcWACC = function(Rt, Ri, Rf, Rcs, Beta, Debt, Equity) {
  # Calculates the annual nominal (adjusted for inflation) weighted average
  # cost of capital of an investment.
  # Rt = Corporate tax rate
  # Ri = Interest rate on debt
  # Rf = Risk free bond rate. Typically 6.4% on government long-term bonds.
  # Rcs = Historical return on common stock.  Typically 7% (1926-1995).
  # Beta = Equity beta
  # Debt = Outstanding leverage
  # Equity = Market value of equity

  wacc = ((1 - Rt) * Ri * Debt + (Rf + (Beta * Rcs)) * Equity) / (Debt +
                                                                Equity)
}
```

```
# Black-Scholes Option Value function
CalcOptionVal = function(P, S, irf, vol, t, sel) {
  # P = current price of the asset
  # S = strike price of the option
  # irf = risk-free interest rate
  # vol = annual volatility
  # t = time to expiration
  # sel = a flag to indicate whether the option considered is a put
  # or a call.
  # call = 1; put = 0

  sel = 1 - selector
  d1 = (log(P / S) + (irf + (vol ^ 2) / 2) * t) / (vol * sqrt(t))
  d2 = d1 - (vol * sqrt(t))
  N1 = pnorm(d1 * (-1) ^ (sel), 0 , 1)
  N2 = pnorm(d2 * (-1) ^ (sel), 0 , 1)

  option.val = (-1) ^ (sel) * (P * N1 - S * exp(-irf * t) * N2)
}
```

Decision Hierarchy and Strategy Table Templates

The following provides a simplified set of instructions for using the integrated decision hierarchy and strategy table templates found in the Excel spreadsheet StrategyTable.xlsx. The spreadsheet is available to download at this book's product page at www.apress.com/9781484234945 by clicking the source code button there.

Decision Hierarchy Worksheet

The decision hierarchy is the tool used to identify and partition decision types. Its first effect is to reduce the number of decision issues to consider.

Decision Issues

Begin the process by capturing all the decision questions the decision team might have. These usually take the form of "Should we do X?" or "Where or how should we do X?"

1. Elicit and list decision issues in the Decision Issues column.

2. Discuss how each issue should be regarded, either as Policy, Strategic (or Open), or Tactical. Policy decisions are foregone decisions or constraints that are (usually) not open for negotiation. Tactical decisions are those that can be deferred because attending to them now will not materially affect the outcome. The remaining decisions are open for further consideration.

© Robert D. Brown III 2018
R. D. Brown III, *Business Case Analysis with R*, https://doi.org/10.1007/978-1-4842-3495-2

3. Categorize each issue with a choice from the drop-down list to the right of the issues in the Type column.

4. Select the array of Decision Issues and corresponding Type, then sort (Data/Sort) the array by the Type column in alphabetical order.

Decision Hierarchy

Once the decision hierarchy is populated appropriately, continue refining the decision team's thoughts about the alternatives that can be exercised for each open decision and how they should be defined.

1. Copy the Policy issues to the Policy list. If the issues were originally posed as questions, you will want to restate them as imperatives.

2. Copy the Strategic issues to the Strategic (Focus) Decisions list. Again, if the issues were originally posed as questions, you will want to distill them to succinct names for categories.

3. Copy the Tactical issues to the Tactical list.

4. For each Strategic Decision, list the alternatives that are conceptually available to the right of a given decision. Each set of alternatives should be exhaustive within the context of the issue at hand and mutually exclusive. If alternatives can be exercised together, combine them as if that combination were an alternative. For example, if Geography is a decision category with initial alternatives of {United States, Europe, Asia, Middle East}, but it makes sense that United States and Europe could be an exclusive alternative, modify the list to {United States, Europe, Asia, Middle East, United States & Europe}. On the other hand, if you would never exercise United States or Europe exclusive of the other, make your list look like {Asia, Middle East, United States & Europe}.

The Decision Hierarchy column can be cut and pasted into other communication documents to clarify how the decision issues were partitioned.

Strategy Table Worksheet

The strategy table is the working palette used by the decision team to consider how decision alternatives might work together in a thematic way to achieve the team's desired goals and objectives.

Strategy Table

Using the strategy table produces a manageable set of decision strategies that significantly reduces the possibly incomprehensibly large number of ways the decision alternatives can be combined and to delimit the combinations that would not make sense in the context of the current decision problem.

1. Start with the Momentum strategy. Copy and paste its colored marker to decision alternatives under each decision category such that the combination of alternatives represents the strategy.

2. Repeat this process for meaningful combinations of alternatives, giving each combination a descriptive, thematic name in the Strategic Theme column. Each theme should be associated with a distinctive colored marker.

3. Let themes range from conservative (mild) in nature to aggressive or even dumb (wild).

Simplified Strategy Table

The simplified strategy table summarizes the working content of the strategy table into a visually clear representation of the decision team's hypotheses for how to create value.

1. For each Strategic theme, select the decision alternative in the drop-down list associated with the colored marker from the earlier table.

2. Fill the row for each strategic theme with the color associated with its marker if helps to make the rows visually distinctive.

Strategy Rationales Worksheet

The strategy rationales worksheet provides a concise way to refer to the meaning of decision strategies and to communicate about them succinctly with other interested parties. For each Strategic Theme, complete a table with the following information:

1. *Name*: Simply provide the name of the strategy.

2. *Description*: Provide a short rationale for why this strategy is being considered and what its key theme addresses.

3. *Objectives*: Describe the compelling goals and objectives that might be obtained.

4. *Benefits*: Name the most compelling benefits that you might obtain by taking this strategy.

5. *Risks*: Name the most compelling undesirable outcomes you might experience by taking this strategy.

6. *Wins if*: What intermediate events must come together to make this strategy succeed?

7. *Loses if*: What intermediate events could come together to make this strategy fail?

8. *Hunch*: How does the decision team think this strategy will play out if actually implemented?

You can use this information to compare with your final decision to see how biased or insightful your thinking might have been originally.

APPENDIX E

VOI Code Samples

Preliminary VOI Example

```
# Mean NPVs corresponding to the p10, p50, p90 quantile values of the
# conditional uncertainty.
dec.val.A <- c(140, 170, 230)
dec.val.B <- c(120, 150, 210)

# The expected NPV of each pathway is the probability weighted average NPV.
# Using the extended Swanson-Megill weights for p10, p50, p90 quantiles.
prob.wts <- c(0.3, 0.4, 0.3)
dec.val.A.m <- sum(dec.val.A * prob.wts)
dec.val.B.m <- sum(dec.val.B * prob.wts)

probs.branch.matrix <- as.array(prob.wts %*% t(prob.wts))

dec.val.A.matrix <- array(rep(dec.val.A, 3), dim = c(3, 3))
dec.val.B.matrix <- t(array(rep(dec.val.B, 3), dim = c(3, 3)))
path.max <- pmax(dec.val.A.matrix, dec.val.B.matrix)
dec.val.prior.info <- sum(path.max * probs.branch.matrix)

voi <- dec.val.prior.info - max(dec.val.A.m, dec.val.B.m)
```

VOI R Scripts

Place the following R script files in each section in a directory, like this:

```
/Applications/R/RProjects/Value of Information
    /Functions.R
    /Assumptions.R
```

© Robert D. Brown III 2018
R. D. Brown III, *Business Case Analysis with R*, https://doi.org/10.1007/978-1-4842-3495-2

```
/Business_Decision_Model.R
/Sensitivity_Analysis.R
/Value_of_Information_1.R
/Value of Information_2.R
```

Run the following scripts in the given order during the same R session:

1. Business_Decision_Model.R

2. Sensitivity_Analysis.R

3. Value_of_Information_1.R (Coarse)

4. Value of Information_2.R (Fine)

Functions

```
CalcBrownJohnson <- function(minlim=-Inf, p10, p50, p90, maxlim=Inf,
samples) {
# This function simulates a distribution from three expert estimates
# for the 80th percentile probability interval of a predicted outcome.
# The user specifies the three parameters and the number of samples.
# The user can also enter optional minimum and maximum limits that
# represent constraints imposed by the system being modeled. These are
# set to -inf and inf, respectively, by default. The process of
# simulation is simple Monte Carlo with 100 samples by default.

    # Create a uniform variate sample space in the interval (0,1).
    U <- runif(samples, 0, 1)
    lenU <- length(U) Create an index in the interval (1,samples) with
    # samples members.
    Uindex <- 1:lenU Calculates the virtual tails of the distribution

    # given the p10, p50, p90 inputs.
    p0 <- pmax(minlim, p50 - 2.5 * (p50 - p10))
    p100 <- pmin(maxlim, p50 + 2.5 * (p90 - p50))
```

```
# This next section finds the linear coefficients of the system of linear
# equations that describe the linear spline, using... [C](A) = (X)
(A) = [C]^-1* (X) In this case, the elements of (C) are found using
# the values (0, 0.1,
# 0.5, 0.9, 1) at the endpoints of each spline segment. The elements
of (X) correspond to the values of (p0, p10, p10, p50,p50, p90,
# p90, p100). Solving for this system of linear equations gives
# linear coefficients that transform values in U to intermediate
# values in X. Because there are four segments inthe linear spline,
# and each segment contains two unknowns, a total of eight
# equations are required to solve the system.

# The spline knot values in the X domain.
knot_vector <- c(p0, p10, p10, p50, p50, p90, p90, p100)

# The solutions to the eight equations at the knot points required to
describe the linear system.
coeff_vals <- c(0, 1, 0, 0, 0, 0, 0, 0,
                0.1, 1, 0, 0, 0, 0, 0, 0,
                0, 0, 0.1, 1, 0, 0, 0, 0,
                0, 0, 0.5, 1, 0, 0, 0, 0,
                0, 0, 0, 0, 0.5, 1, 0, 0,
                0, 0, 0, 0, 0.9, 1, 0, 0,
                0, 0, 0, 0, 0, 0, 0.9, 1,
                0, 0, 0, 0, 0, 0, 1, 1)

# The coefficient matrix created from the prior vector. It looks like
the following matrix:
            # [,1] [,2] [,3] [,4] [,5] [,6] [,7] [,8]
  # [1,]  0.0   1  0.0   0  0.0   0  0.0   0
  # [2,]  0.1   1  0.0   0  0.0   0  0.0   0
  # [3,]  0.0   0  0.1   1  0.0   0  0.0   0
  # [4,]  0.0   0  0.5   1  0.0   0  0.0   0
  # [5,]  0.0   0  0.0   0  0.5   1  0.0   0
  # [6,]  0.0   0  0.0   0  0.9   1  0.0   0
  # [7,]  0.0   0  0.0   0  0.0   0  0.9   1
  # [8,]  0.0   0  0.0   0  0.0   0  1.0   1
```

```
    coeff_matrix <- t(matrix(coeff_vals, nrow=8, ncol=8))

    #The inverse of the coefficient matrix.
    inv_coeff_matrix <- solve(coeff_matrix)

    #The solution vector of the linear coefficients.
    sol_vect <- inv_coeff_matrix %*% knot_vector

    X = (U <= 0.1) * (sol_vect[1, 1] * U + sol_vect[2, 1]) +
        (U > 0.1 & U <= 0.5) * (sol_vect[3, 1] * U + sol_vect[4, 1]) +
        (U > 0.5 & U <= 0.9) * (sol_vect[5, 1] * U + sol_vect[6, 1]) +
        (U > 0.9 & U <= 1) * (sol_vect[7, 1] * U + sol_vect[8, 1])

    return(X)
}

CalcScurv <- function(y0, t, tp, k = 0) {
  # This function models the sigmoid 1 / (1 + exp(-g*t)) with four analytic
  # parameters.
  # y0 = saturation in first period.
  # t = the index along which the s-curve responds, usually thought of as
time.
  # k = the offset in t for when the sigmoid begins. Subtract k for a right
  #     shift. Add k for a left shift.
  # tp = the t at which s-curve achieves 1-y0.

  this.s.curve = 1 / (1 + (y0 / (1 - y0)) ^ (2 * (t + k) / tp - 1))
  return(this.s.curve)
}

CalcBizModel <- function(Time, N, pus, ttp, p, sga, cogs, tr, i, dr) {
  # The function that represents the business model reflected by the
    influence diagram.

    # Time = the time index
    # N = the number of simulation samples
    # pus = peak units sold
    # ttp = time to peak units sold
    # p = price, $/unit
```

```
# sga = sales, general, and admin, % revenue
# cogs = cost of goods sold, $/unit
# tr = tax rate, %
# i = initial investment, $
# dr = discount rate, %/year

init <- t(array(0, dim=c(length(Time), N)))
ann.units.sold <- init
revenue <- init
profit <- init
tax <- init
cash.flow <- init
npv <- rep(0, N)
samp.index <- 1:N
for (s in samp.index) {
  # annual units sold
  ann.units.sold[s, ] <-
    (Time > min(Time)) * pus[s] * CalcScurv(0.02, Time, ttp[s], -1)

  # annual period revenue
  revenue[s, ] <- p[s] * ann.units.sold[s, ]

  # annual period profit
  profit[s, ] <-
    revenue[s, ] - (sga[s] * revenue[s, ]) - (cogs[s] * ann.units.
    sold[s, ])

  # annual period tax
  tax[s, ] <- tr * profit[s, ]

  # annual period cash flow
  cash.flow[s, ] <-
    (Time > min(Time)) * (profit[s, ] - tax[s, ]) - (Time == min
    (Time)) * i[s]

  # net present value of the annual period cash flow
  npv[s] <- sum(cash.flow[s, ] / (1 + dr) ^ Time)
}
```

```
    # Collect all intermediate calculations in a list to be used for other
    # calculations or reporting.
    calc.vals <- list(ann.units.sold = ann.units.sold,
                        revenue = revenue,
                        profit = profit,
                        tax = tax,
                        cash.flow = cash.flow,
                        npv = npv
                    )
    return(calc.vals)
}

CalcModelSensitivity <- function(unc.list, sens.q) {
  # unc.list = A list that contains the uncertain variables' samples used in
  #            the business model.
  # sens.q = a vector that contains sensitivity test quantiles

  # Create an index from 1 to the number of uncertainties used in the business
  # model.
  unc.index <- 1:length(unc.list)

  # Assign the values of the uncertainties list to a temporary list
  uncs.temp <- unc.list

  # Initialize a table to contain mean NPVs of the business model as each
  # uncertainty is set to the sensitivity quantile values.
  sens.table <-
    array(0, dim = c(length(unc.list), length(sens.q)))
  row.names(sens.table) <- names(unc.list)
  colnames(sens.table) <- sens.q

  for (u in unc.index) {
    for (s in 1:length(sens.q)) {

      # Iterate across the uncertainties and elements of the sensitivity
      # quantile values and temporarily replace each uncertainty's samples with
      # the uncertainty's value at each quantile value.
      uncs.temp[[u]] <- rep(quantile(unc.list[[u]], sens.q[s]), samps)
```

```
    # Populate the sensitivity table with the mean NPV values calculated in
    # the business model using the values in the temporary uncertainty list.
    sens.table[u, s] <- mean(
      CalcBizModel(
        time,
        samps,
        uncs.temp$peak.units.sold,
        uncs.temp$time.to.peak,
        uncs.temp$price,
        uncs.temp$sga,
        uncs.temp$cogs,
        tax.rate,
        uncs.temp$investment,
        disc.rate
      )$npv
    )
  }
  # Reset the temporary uncertainty list back to the original uncertainty
  list.
  uncs.temp <- unc.list
}
return(sens.table)
}
```

Assumptions

```
# time horizon of the model
time.horizon <- 10 # yr
time <- 0:time.horizon

# Ensure reproducible results
set.seed(98)

# number of samples used for the simulation.
samps <- 30000
```

```
tax.rate <- 0.35 # %
disc.rate <- 0.10 # %/yr

# Uncertainties have the following units.
# peak.units.sold, max units/year after year 0
# time.to.peak, years
# price, $/unit
# cogs, $/unit
# sga, % revenue
# investment, $ in year 0

# uncertainty parameters for decision 1
dec1.uncs <- list(
  peak.units.sold = round(CalcBrownJohnson(0, 10000, 12500, 18000, , samps)),
  time.to.peak = CalcBrownJohnson(0, 2, 3, 5, , samps),
  price = CalcBrownJohnson(0, 90, 95, 98, , samps),
  cogs = CalcBrownJohnson(0, 15, 17, 20, , samps),
  sga = CalcBrownJohnson(0, 0.1, 0.12, 0.14, 1, samps),
  investment = CalcBrownJohnson(0, 100000, 115000, 150000, , samps)
)

# uncertainty parameters for decision 2
dec2.uncs = list(
  peak.units.sold = round(CalcBrownJohnson(0, 9000, 17000, 25000, , samps)),
  time.to.peak = CalcBrownJohnson(0, 2, 4, 6, , samps),
  price = CalcBrownJohnson(0, 95, 100, 103, , samps),
  cogs = CalcBrownJohnson(0, 16, 18, 21, , samps),
  sga = CalcBrownJohnson(0, 0.1, 0.13, 0.16, 1, samps),
  investment = CalcBrownJohnson(0, 135000, 155000, 200000, , samps)
)
```

Business Decision Model

```
# Import function and data source files.
source("/Applications/R/RProjects/Value of Information/Functions.R")
source("/Applications/R/RProjects/Value of Information/Assumptions.R")
```

```
# Run the business decision models and return sample outputs of NPV.
bdm1 <-
  CalcBizModel(
    time,
    samps,
    dec1.uncs$peak.units.sold,
    dec1.uncs$time.to.peak,
    dec1.uncs$price,
    dec1.uncs$sga,
    dec1.uncs$cogs,
    tax.rate,
    dec1.uncs$investment,
    disc.rate
  )$npv / 1e6

bdm2 <-
  CalcBizModel(
    time,
    samps,
    dec2.uncs$peak.units.sold,
    dec2.uncs$time.to.peak,
    dec2.uncs$price,
    dec2.uncs$sga,
    dec2.uncs$cogs,
    tax.rate,
    dec2.uncs$investment,
    disc.rate
  )$npv / 1e6

# Calculate the mean NPV of each decision and their difference.
m.bdm1 <- mean(bdm1)
m.bdm2 <- mean(bdm2)
diff.bdm <- mean(bdm2 - bdm1)

# Tabulate the average model results.
m.bdm <- array(c(m.bdm1, m.bdm2), dim=c(2, 1))
colnames(m.bdm) <- "Mean NPV of decision [$M]"
```

269

```r
rownames(m.bdm) <- c("Decision1", "Decision2")
print(signif(m.bdm,3))
print(paste("The difference in mean value between strategies:", "$",
            signif(diff.bdm, 3), "M"))

# Plot the cumulative probability distribution of the NPVs.
par(mar = c(12, 5, 5, 2) + .05, xpd = TRUE)
legend.idents <- c("Decision 1", "Decision 2")
plot.ecdf(
  bdm1,
  do.points = TRUE,
  col = "red",
  main = "NPV of Business Decision",
  xlab = "NPV [$M]",
  ylab = "Cumulative Probability",
  xlim = c(min(bdm1, bdm2), max(bdm1, bdm2)),
  tck = 1
)
plot.ecdf(
  bdm2,
  do.points = TRUE,
  add = TRUE,
  col = "blue")
legend(
  "bottom",
  inset = c(0, -.45),
  legend = legend.idents,
  text.width = 2.5,
  ncol = 2,
  pch = c(13, 14),
  col = c("red", "blue")
)
```

Sensitivity Analysis

```
# time -> the time index, year
# peak units sold -> uncertainty
# time to peak units sold -> uncertainty, years
# price -> uncertainty, $/unit
# sales, general, and admin -> uncertainty, % revenue
# cost of goods sold -> uncertainty, $/unit
# tax rate -> % profit
# initial investment -> uncertainty, $
# discount rate -> %/year

# Create a vector of values representing the sensitivity quantiles.
sens.range <- c(0.1, 0.5, 0.9)

npv.sens1 <- CalcModelSensitivity(
  unc.list = dec1.uncs,
  sens.q = sens.range
) / 1e6

npv.sens2 <- CalcModelSensitivity(
  unc.list = dec2.uncs,
  sens.q = sens.range
) / 1e6

print("Decision 1 Sensitivity to Uncertainty [$M]")
print(signif(npv.sens1[, c(1, length(sens.range))], 3))

print("Decision 2 Sensitivity to Uncertainty [$M]")
print(signif(npv.sens2[, c(1, length(sens.range))], 3))

# Find the rank order of the uncertainties by the declining range of variation
# they cause in the NPV. Base this on the decision with the highest mean value.
mean.bdm.list <- c(m.bdm1, m.bdm2)
bdm.sensitivity.list <- list(npv.sens1, npv.sens2)
npv.sens.rank <-
  order(abs(bdm.sensitivity.list[[which.max(mean.bdm.list)]][, 1] -
            bdm.sensitivity.list[[which.max(mean.bdm.list)]][, length
            (sens.range)]),
        decreasing = FALSE)
```

```
# Reorder the variables in the sensitivity arrays and names array by the rank
# order.
ranked.npv.sens2 <- npv.sens2[npv.sens.rank, c(1, length(sens.range))]
ranked.npv.sens1 <- npv.sens1[npv.sens.rank, c(1, length(sens.range))]

# Plot the tornado chart for Decision 2
par(mai = c(1, 1.75, .5, .5))
barplot(
  t(ranked.npv.sens2) - m.bdm2,
  main = "NPV Sensitivity to Uncertainty Ranges",
  names.arg = names(ranked.npv.sens2),
  col = "blue",
  xlim = c(min(npv.sens1, npv.sens2), max(npv.sens1, npv.sens2)),
  xlab = "Decision2 NPV [$M]",
  beside = TRUE,
  horiz = TRUE,
  offset = m.bdm2,
  las = 1,
  space = c(-1, 1),
  cex.names = 1
)

# Plot the tornado chart for Decision 1
par(mai = c(1, 1.75, .5, .5))
barplot(
  t(ranked.npv.sens1) - m.bdm1,
  main = "NPV Sensitivity to Uncertainty Ranges",
  names.arg = names(ranked.npv.sens1),
  col = "red",
  xlim = c(min(npv.sens1, npv.sens2), max(npv.sens1, npv.sens2)),
  xlab = "Decision1 NPV [$M]",
  beside = TRUE,
  horiz = TRUE,
```

```
  offset = m.bdm1,
  las = 1,
  space = c(-1, 1),
  cex.names = 1
)
```

Value of Information 1: Coarse

```
# Define a vector of the extended Swanson-Megill probability weights. These
# correspond to the values in the sens.range vector.
prob.wts <- c(0.3, 0.4, 0.3)

# Create the 3x3 matrix of the products of the uncertainty branch
  probabilities.
probs.branch.matrix <- prob.wts %*% t(prob.wts)

# Typically, we run a business model for each decision with critical uncertainty
# peak.units.sold set to its respective p10, p50, p90 values. Find the mean NPV
# of the business model at each of these points. However, we already did
  this in
# the Sensitivity_Analysis.R script, so all we need to do is use the slice of
# the sensitivity table associated with the peak.units.sold row.
bdm1.sens <- npv.sens1[1, ]
bdm2.sens <- npv.sens2[1, ]

# Create a 3x3 matrix of the values in the bdm1.sens and bdm2.sens vectors.
# Transpose the values in the second matrix
bdm1.sens.matrix <- array(rep(bdm1.sens, 3), dim = c(3, 3))
bdm2.sens.matrix <- t(array(rep(bdm2.sens, 3), dim = c(3, 3)))

# Find the parallel maximum value between these matrices. This represents
# knowing the maximum value given the prior information about the combinations
# of the outcomes of each uncertainty.
bdm.sens.prior.info <- pmax(bdm1.sens.matrix, bdm2.sens.matrix)

# Find the expected value of the matrix that contains the decision values with
# prior information. This is the value of knowing the outcome before making a
```

```
# decision.
bdm.prior.info <- sum(bdm.sens.prior.info * probs.branch.matrix)

# Find the maximum expected decison value without prior information.
bdm.max.curr.info <- max(mean(bdm1), mean(bdm2))

# The value of information is the net value of knowing
# the outcome beforehand compared to the decison with
# the highest value before the outcome is known.
val.info <- (bdm.prior.info - bdm.max.curr.info)

print("Decision Value [$M]")
value.report <- list("Prior Information" = signif(bdm.prior.info, 3),
                     "Current Information" = signif(bdm.max.curr.info, 3),
                     "Value of Information" = signif(val.info, 3))
print(value.report)
```

Value of Information 2: Fine

```
# Find the histogram of the critical uncertainty for each decision pathway.
dec1.branch.hist <- hist(dec1.uncs$peak.units.sold, plot = TRUE)
dec2.branch.hist <- hist(dec2.uncs$peak.units.sold, plot = TRUE)

# Find the midpoints of the bins.
dec1.branch.bins <- dec1.branch.hist$mids
dec2.branch.bins <- dec2.branch.hist$mids

# Find the frequency of the bins in the histograms.
dec1.branch.cnts <- dec1.branch.hist$counts
dec1.branch.probs <- dec1.branch.cnts / samps

dec2.branch.cnts <- dec2.branch.hist$counts
dec2.branch.probs <- dec2.branch.cnts / samps

# Find the sensitivity of the NPV to the midpoints of the bins in each
# uncertainty.
bdm1.sens <- c(0)
```

```
for (b in 1:length(dec1.branch.bins)) {
  bdm1.sens[b] <- mean(
    CalcBizModel(
      time,
      samps,
      rep(dec1.branch.bins[b], samps),
      dec1.uncs$time.to.peak,
      dec1.uncs$price,
      dec1.uncs$sga,
      dec1.uncs$cogs,
      tax.rate,
      dec1.uncs$investment,
      disc.rate
    )$npv / 1e6
  )
}

bdm2.sens <- c(0)
for (b in 1:length(dec2.branch.bins)) {
  bdm2.sens[b] <- mean(
    CalcBizModel(
      time,
      samps,
      rep(dec2.branch.bins[b], samps),
      dec2.uncs$time.to.peak,
      dec2.uncs$price,
      dec2.uncs$sga,
      dec2.uncs$cogs,
      tax.rate,
      dec2.uncs$investment,
      disc.rate
    )$npv / 1e6
  )
}
```

```r
# Create the MxN matrix of the products of the uncertainty branch probabilities.
probs.branch.matrix <- dec1.branch.probs %*% t(dec2.branch.probs)

# Create an MxN matrix of the values in the bdm1.sens and bdm2.sens vectors.
# Transpose the values in the second matrix
bdm1.sens.matrix <-
  array(rep(bdm1.sens, length(dec2.branch.probs)),
        dim = c(length(dec1.branch.probs), length(dec2.branch.probs)))

bdm2.sens.matrix <-
  t(array(rep(bdm2.sens, length(dec1.branch.probs)),
          dim = c(length(dec2.branch.probs), length(dec1.branch.probs)
  )))

# Find the parallel maximum value between these matrices. This represents
# knowing the maximum value given the prior information about the combinations
# of the outcomes of each uncertainty.
bdm.sens.prior.info <- pmax(bdm1.sens.matrix, bdm2.sens.matrix)

# Find the expected value of the matrix that contains the decision values with
# prior information. This is the value of knowing the outcome before making a
# decision.
bdm.prior.info <- sum(bdm.sens.prior.info * probs.branch.matrix)

# Find the maximum expected decison value without prior information.
bdm.max.curr.info <- max(mean(bdm1), mean(bdm2))

# The value of information is the net value of knowing
# the outcome beforehand compared to the decison with
# the highest value before the outcome is known.
val.info <- (bdm.prior.info - bdm.max.curr.info)

print("Decision Value [$M]")
value.report <- list("Prior Information" = signif(bdm.prior.info, 3),
                     "Current Information" = signif(bdm.max.curr.info, 3),
                     "Value of Information" = signif(val.info, 3))
print(value.report)
```

Index

A

Analysis
   business decision makers, 11
   canonical influence diagrams, 14
   deterministic base case, 9–10
   deterministic sensitivity (*see*
      Deterministic sensitivity analysis)
   file structure, 15–16
   financial model (*see* Deterministic
      financial model)
   influence diagram, 12–13
   intermediate values, 14
   pro forma table, 38–40
   project management efforts, 14
   risk layer, 10
   seminars, quantitative business, 9
   style guide, 16–17

B

Biases
   anchoring, 165
   availability, 165
   bandwagon bias/groupthink, 165
   blind spot, 165
   cognitive and motivational, 167
   confirmation, 165
   elicitation process, 166
   entitlement, 165
   expert overconfidence, 166
   false precision, 166
   frame, 166
   incentives, 166
   sand bagging, 166
   selection, 166
   unwarranted optimism, 166
BrownJohnson distribution
   business case simulations, 59
   cumulative probability, 63
   function, 249–256
   linear spline approach, 60–62
   p0 and p100 values, 60
   parametric guidance, 60
   SME, 60

C

CalcBizModel, 188, 191–192, 194
CalcBrownJohnson, 188, 190
CalcModelSensitivity, 188, 192–193
CalcScurv, 188, 190
Capital expense (CAPEX), 18
   apply() function, 25
   capital costs, 24
   depr values, 26
   depreciation period, 23
   depr.matrix, 25
   for loop, 22, 23, 25
   gross profit, taxable income, 22

© Robert D. Brown III 2018
R. D. Brown III, *Business Case Analysis with R*, https://doi.org/10.1007/978-1-4842-3495-2

Get the eBook for only $5!

Why limit yourself?

With most of our titles available in both PDF and ePUB format, you can access your content wherever and however you wish—on your PC, phone, tablet, or reader.

Since you've purchased this print book, we are happy to offer you the eBook for just $5.

To learn more, go to http://www.apress.com/companion or contact support@apress.com.

Apress®

Printed in the United States
By Bookmasters